Measuring America

WALKER & COMPANY
NEW YORK

Measuring America

HOW AN UNTAMED WILDERNESS
SHAPED THE UNITED STATES
AND FULFILLED THE PROMISE OF DEMOCRACY

Andro Linklater

First published in the United States of America in 2002
by Walker Publishing Company, Inc.

Published simultaneously in Canada by Fitzhenry and Whiteside,
Markham, Ontario L3R4T8

For information about permission to reproduce selections from this book, write to
Permissions, Walker & Company, 435 Hudson Street, New York, New York 10014

Every effort has been made to locate and contact all the
holders of copyright to material reproduced in this book.

Library of Congress Cataloging-in-Publication Data
Linklater, Andro.
Measuring America : how an untamed wilderness shaped the United States and fulfilled the
promise of democracy / Andro Linklater.
    p.   cm.
Includes bibliographical references and index.
ISBN 0-8027-1396-3
1. United States—Geography. 2. United States—Surveys—History. 3. United States—
Territorial expansion. 4. National characteristics, American. 5. Public lands—
United States—History. 6. Frontier and pioneer life—United States. 7. Ohio River
Valley—Geography. 8. Ohio River Valley—Surveys—History. 9. Surveying—
United States—History. 10. Surveyors—United States—History. I. Title.
                                E161.3 .L46   2002
  973—dc21   2002073573

Visit Walker & Company's Web site at www.walkerbooks.com

Art Credits: Pages ii-iii: map of Kentucky by John Filson, published in 1793, used courtesy of Art Lassagne, Gold Bug Historic Maps and Software, www.goldbug.com. Pages 3, 37, and 43: New York Public Library. Page 15: Science Museum/Science and Society Picture Library. Page 17: Museum of Surveying. Pages 28, 65, 95, 114, 129, 158, 165, 173, 177, 178-79, 182, 186, 195, 208, and 232: The Library of Congress. Page 46: Collection of The British Library. Page 52: Collection of The New York Historical Society. Page 56: National Portrait Gallery, Smithsonian Institution, Bequest of Charles Francis Adams. Pages 100 and 101: National Maritime Museum, Greenwich, England. Page 107: Independence National Historical Park. Page 134: Biblioteque Centrale du Musée National d'Histoire Naturelle. Pages 139, 204, and 205: National Institute of Standards and Technology Museum. Page 147: The Great Western Reserve Historical Society. Page 151: The Georgia State Archives. Page 184: Wichita State University Special Collections. Page 211: Museum of Surveying. Page 217: Denver Public Library Western History Department. Page 224: United States Geological Survey. Page 262: Author's collection.

Book design by Ralph Fowler

Printed in the United States of America

2 4 6 8 10 9 7 5 3 1

"God did make the world to be inhabited by mankind, and to have his name known to all nations, and from generation to generation as the people increased and dispersed themselves into such countries as they found most convenient. And there in Florida, Virginia, New England and Canada is more land than all the people of Christendom can cultivate, and yet more to spare than all the natives of those countries can use and cultivate. And shall we here keep such a small island, and at such great rents and rates, where there is so much of the world uninhabited, and as much more in other places, and as good or rather better than any we now possess, were it cultivated and used accordingly?" —John Smith, *Advertismentz for the Inexperienced Planters of New England,* Haviland, London, 1631

"The great and chief end, therefore, of men's uniting into commonwealths, and putting themselves under government, is the preservation of their property." —John Locke, *An Essay concerning the true, original extent and end of civil Government,* London, 1690

CONTENTS

Introduction

EAST LIVERPOOL, OHIO, sits on the banks of the Ohio River just outside the Pennsylvania border. In every country with an industrial history there are towns like this. They were built close to the coalfields that provided their energy and on the banks of large rivers that carried away their heavy products. On the Clyde in Scotland, the Tees in England, the Ruhr in Germany, mighty structures of iron and steel were smelted, beaten, and annealed to make the skeleton for the first modern society, and the towns grew rich and confident in a gray haze of fumes and steam and effluent. Today their colors are rusty metal and faded red brick, the air is clear, and poverty creates uncertainty.

Clay was the material that made East Liverpool wealthy, Pennsylvania coal fueled the furnaces, and the Ohio River carried away enough plates and cups to let the inhabitants boast that it was "the pottery capital of the world." Where once dozens of factories and chemical works lined the banks, the only reminder now of its industrial past is a solitary chimney stack pumping heavy coils of smoke into the sky. The barges still come up the Ohio River carrying minerals and fertilizers but mostly for use across the river in the nearby Pittsburgh area. They unload at depots like that belonging to the S. H. Bell Company, whose functional warehouses and concrete docks annually handle thousands of tons of steel and copper at the upriver end of East Liverpool.

On the road above the Bell Company's dock, Pennsylvania Route 68 invisibly changes to Ohio Route 38, and trees half hide some signs by the roadside. The place could hardly be more anonymous. Even someone familiar with the historical significance of this particular spot, who has traveled several thousand miles to find it, and whose eyes are flickering wildly from the narrow blacktop to the grassy verge between road and river, can drive a couple of hundred yards past it before hitting the brakes.

The language of the signs is equally undemonstrative. A stone marker carries a plaque headed "The Point of Beginning" that reads, "1112 feet south of this spot was the point of beginning for surveying the public lands of the United States. There on September 30, 1785, Thomas Hutchins, first Geographer of the United States, began the Geographer's Line of the Seven Ranges."

There is nothing to suggest that it was here that the United States began to take physical shape, nothing to indicate that from here a grid was laid out across the land that would stretch west to the Pacific Ocean, and north to Canada, and south to the Mexican border, and would cover more than three million square miles, and would create a structure of landownership unique in history, and would provide the invisible web that supported the legend of the frontier with its covered wagons and cowboys, its farmers and gold miners, and would insidiously permeate its formation into the unconscious mind of every American who ever owned a square yard of soil.

It is hilly country, covered in the same oak, dogwood, and hickory that Hutchins saw, and in the bright light of a September afternoon it rises high above the broad river in crimson and copper waves. "For the distance of 46 chains and 86 links West," Hutchins wrote in his first description of the territory, "the Land is remarkably rich with a deep, black Mould, free from Stone." He was Robinson Crusoe, landed in an uncharted wilderness, and his purpose was to measure it so that it could be sold.

It is easy to miss the significance of what he proposed to do. Surveying is old—its 5,000-year history originates in riverside communities in the Middle East and in Egypt who needed to mark the boundaries of areas to be irrigated by the annual flooding of the Tigris and the Nile. Measurement is older still, and the process is so universal as not to merit a second thought. Yet without it,

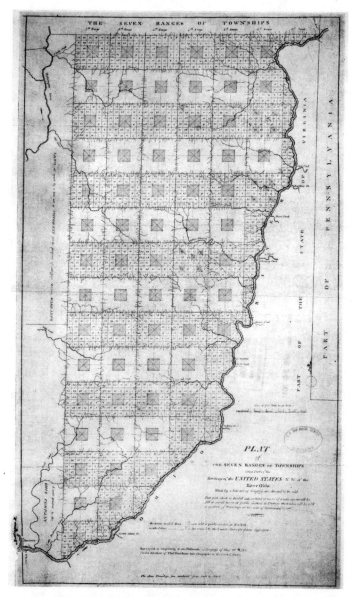

The Point of Beginning (uppermost right boundary).

the exchange of goods and services cannot take place. Measuring a length of cloth or a herd of animals or a day's labor gives it a value in terms of size, or number, or achievement, which enables others in the community to offer something else—food, protection, even love or loyalty—of equivalent value. Consequently, it is almost as necessary to human society as language, and it occurs in the earliest civilizations, long before the development of writing.

Incised bone and clay counters used for counting or exchange have been found at Neolithic village sites in Turkey dating from the ninth millennium B.C.E.—more than 10,000 years ago—while the Sumerian cuneiform script, the earliest form of writing, does not appear until 5,000 years later. In a modern society, this remarkable human achievement is almost invisible. Glancing down supermarket shelves, we examine the prices but have to make an effort to register the weight of a cereal box or the capacity of a carton of juice. Weights and measures are a given. A pound or a gallon, like a mile or an acre, will be the same from Florida to Alaska. And so will a bushel of wheat and a cord of wood and a hundred other units of measurement. It is a language that is picked up automatically and spoken without conscious thought.

Only when it changes, when half a gallon of cola became 2 liters in the 1990s, for instance, or a fifth of whiskey reappeared as 750 milliliters, is there a reminder that there is nothing certain about these units after all. They are not a given but an extraordinary construct, and one of the identifying marks of social life. Without the conscious decision to agree on a way of measuring, cooperative activity could hardly take place. With it, marketplaces and increasingly sophisticated economies can develop, matching barter, cash, or credit to whatever is owned by one person and desired by another.

Thus, by measuring out the wilderness, Hutchins would make it possible for someone to buy and own it. This was a revolutionary concept. For centuries the land had been lived in by the Delaware and passed through by the Miami and occupied by the Iroquois, but no one had ever owned it. No one had ever thought of owning it. The idea of one person *owning* land did not yet exist on the west bank of the Ohio.

Indeed, throughout most of the world and most of history, such a possibility was inconceivable. Individuals could certainly own the use of land, whole dynasties might grow up around a particular estate and the right of each generation to occupy and exploit that parcel of land would be unquestioned, but it could not be owned as a house or a bed or a pig was owned. The span of one life was too brief to possess the earth that continued forever. Territory defined the community, or society, or even the nation that lived there. A monarch or a ruler who embodied the nation might own it, but no one else. The idea that land might be treated as property belonging to an individual, to be traded, borrowed against, and speculated on like any other commodity, required a fundamental shift in thinking. The concept had been incubated in Tudor England, and taken shape in colonial America, but only with American independence, achieved just two years before Hutchins's arrival on the Ohio, had it been fully realized. This was the magic that Hutchins would introduce into the western lands, the transformation of the wilderness into property.

The wand that would make it possible was there in the first sentence of his report. The land would be measured in chains and links. In most circumstances, a chain imprisons; here it released. What it released from the billowing, uncharted land was a single element—a distance of 22 yards. That was the length of the chain. Invented by the seventeenth-century English mathematician Edmund Gunter, it was the surveyor's one indispensable tool, and the fact that its 22 yards were to become integral not only to the game of cricket in his own country (it is the length of the pitch), but to the town-planning of almost every major city in the United States (the lengths of most city blocks are multiples of it), was a tribute to the instrument's usefulness. Repeated often enough, added, squared, and multiplied, measurement with Gunter's chain would give to the land beyond the Ohio a numerical value that someone could compute in money.

But there was a double significance to Hutchins's work. Like all other units of measurement then in use, the length of Gunter's chain was ultimately derived from human activity and the dimensions of the human body. Hutchins's survey began at a critical moment in the history of ideas, when for the first time in 10,000 years those traditional measurements were challenged by systems derived from scientific discoveries about gravity and the size of the earth. In France and Britain, scientists were devising new decimal units, from which the metric system would shortly emerge, but they lagged behind the United States. Before Hutchins crossed the Ohio, Thomas Jefferson had already invented a decimal system based on the length of the equator, which he insisted should be used in the survey. Nor was he alone in arguing for new, scientific, decimal meaures. Of the founding fathers, George Washington, James Madison, James Monroe, and even Alexander Hamilton supported him, and all understood that the system used in the wilderness would eventually become the system used by the entire United States.

Thus, what began at the spot now covered by the Bell Company's concrete dock was not just a survey but the realization of two of the most potent ideas that have shaped America. The first of these was a unique system of measurement that harks back to the dawn of history and, through the grid that the survey laid across America, was to leave its mark on almost every acre of real estate, every farm and every city block west of the Appalachians. The second, and perhaps more important, idea was that any individual could own outright the land that Hutchins and his followers were to measure. Around that revolutionary perception grew a society whose economic system and democratic outlook were unlike any that had come before.

In his poem "The Gift Outright," quoted at the end of this book, Robert Frost grasped at another thought, more powerful still. In the end, owning a parcel of earth can never be quite like owning other forms of property. The land has its own magic, and those who seek to possess it are liable to end up being possessed by it. It was the desire to own this particular land, Frost mused, that made its owners American.

These were the great forces that Geographer to the United States Captain Thomas Hutchins set in motion when he first unrolled the loops of his chain at the Point of Beginning.

ONE

The Invention of Landed Property

THE IMPOSING LIBRARY of the Royal Institution of Chartered Surveyors in London is strategically situated. In one direction its tall windows look over the street to Whitehall, where the Tudor and Stuart sovereigns ruled in the sixteenth and seventeenth centuries, and in another they gaze across Parliament Square toward the House of Commons, power base of the rising class of landed gentry who during those two centuries challenged the royal authority. It is just possible to imagine the atmosphere of righteous indignation and pervading apprehension that accompanied the struggle between the two, but in the small, leather-bound books kept in the institution's library, the reality that gave rise to the battles remains vividly alive.

In the earliest, such as Master John Fitzherbert's *The Art of Husbandry*, published in 1523, a surveyor still filled his original, feudal role as the executive officer of a landed nobleman. His duty was primarily to oversee (the word *surveyor* is derived from the French *sur* [over] and *voir* [see]) the estate. He was to walk over the land and make a note of the boundaries—the "buttes and bounds"—of the tenants' holdings, and then to assist in drawing up the official record or court roll of what duties they owed. A model report, Fitzherbert suggested, would record that the land of a particular tenant "lyeth between the mill on the north side, and the South Field on the south, butteth upon the hyway, and conteyneth xii perches and x fote [feet] in bredthe by the hyway, and ix perches in length, and payeth . . . two hennes at Christmas and two capons at Easter."

To "butt" upon something is to encounter or meet it, for which the equivalent word was *mete*. This ancient method of surveying, which identified the boundary of an estate by the points where it met other boundaries or visible objects, thus became known as "metes and bounds." Even in 1523, English landlords were engaged in a practice that was to transform the feudal order. There were infinite variations in feudalism, but this was its heart—that the land was the state, and only the head of state could own it outright. The dukes and barons, the king's tenants-in-chief, technically held their broad acres *of* the Crown in return for the dues or service they paid, and their vassals held their narrower farms *from* the great lords in return for rent or service, and so on down to the villeins who often had no land at all but exchanged goods, service, or rent for the right to work it. What the sixteenth-century manuals inadvertently reveal, as they detail the surveyor's duties, is how that order was subverted from within.

Under the old system, tenants farmed narrow strips, or rigs, of land, often widely separated so that good and poor soil was distributed evenly among those who actually worked the land. For centuries, impatient land users had attempted to consolidate the strips into single compact fields that could be "enclosed" by a fence or hedge so that crops were not trodden down or herds scattered, but the pattern remained fundamentally intact. In the early sixteenth century, however, a period of savage inflation occurred, and every lord and tenant tried to squeeze the maximum profit from the land. Repeatedly, Fitzherbert stressed the need for the surveyor to realize that enclosed land was more valuable than the strips and common pasture because it could be made more productive. The urgency was unmistakable, yet essentially the old values were still in place.

Then in 1531 came the publication of *The Boke Named the Governour* by Sir Thomas Elyot, which gave advice on the education of rulers and landowners. An essential part of their training was to learn how to draw so that, according to Elyot, they could make a map or "figure" of their estates. In this way, they would have a clear picture of what they owned, or as Elyot put it, "in visiting his own dominions, [the governor] shall set them out in figure, in such wise that his eye shall appear to him where he shall employ his study and treasure." In the course of the sixteenth century, it became a habit of English landowners to have their estates and the surrounding countryside measured and then mapped. By 1609 John Norden could insist in the *Surveior's Dialogue* that "the [map] rightly drawne by true information, describeth so lively an image of a Manor . . . as the Lord sitting in his chayre, may see what he hath, where and how it lyeth, and in whose use and occupation every particular is."

There was a special significance in making this part of the surveyor's duty, because in that era only the rulers of states and cities made maps. A map was a political document. It not only described territory but asserted ownership of it as well. In 1549 a map of Newfoundland and the North Atlantic seaboard detailing Sebastian Cabot's discoveries was displayed in the Privy Gallery at Whitehall outside the royal council chamber, so that foreign ambassadors waiting to see the sovereign would know of England's claims overseas. When the Flemish cartographer Abraham Ortelius produced the first modern atlas in 1570, his *Theatrum orbis terrarum,* with the freshly discovered territories of the New World and the newly explored Pacific and Indian Oceans, he took care to dedicate it to his sovereign, Philip II of Spain, and to ensure that Philip could find in it his own claims across the ocean.

Consequently, when the sixteenth-century English landowners ordered a map of their estate, they were implicitly asserting a claim to ownership in a way that only rulers of states and kingdoms had been able to make. For a long time only landowners in England made such a claim. Surveying manuals were published in the German states, but there were hardly any estate maps until late in the seventeenth century. Sweden produced its first national map in the sixteenth century, but it was 100 years later before noblemen began measuring and mapping their estates. In sixteenth-century France, the Jesuits taught math and all the theory needed by a surveyor, but no plats, or *plans parcellaires,* of aristocratic land were drawn before 1650. The first Spanish maps appeared as early as 1508, but another 200 years passed before it became the custom for Spanish lords to measure their lands. Only in the economically sophisticated Netherlands, where the mathematician Gemma Frisius wrote the first manual on mapmaking, *A Method of De-*

lineating Places, in 1533, were farms, especially those close to cities, measured and mapped, yet even there the aristocrats' landed estates remained feudal. But in England, estate maps were so common that an inventory of Henry VIII's possessions made at his death in 1547 showed he had "a black coffer covered with fustian of Naples [which was] full of plattes."

If there is a single date when the idea of land as private property can be said to have taken hold, it is 1538. In that year a tiny volume was published with a long title that began, "*This boke sheweth the maner of measurynge of all maner of lande . . .*" In it, the author, Sir Richard Benese, described for the first time in English how to calculate the area of a field or an entire estate. He was probably borrowing from Frisius, but his values were purely English. Noting that sellers tended to exaggerate the size of a property whereas buyers were inclined to underestimate it, he advised the surveyor to approach the task in a careful and methodical manner.

"When ye shall measure a piece of any land ye shall go about the boundes of it once or twice, and [then] consider well by viewing it whether ye may measure it in one parcel wholly altogether or else in two or many parcels." Measuring it in "many parcels," he explained, was necessary when the field was an uneven, irregular shape; by dividing it up into smaller, regular shapes like squares and oblongs and triangles it became easy to calculate accurately the total area. The distances were to be carefully measured with a rod or pole, precisely 16½ feet long, or a cord. And finally, the surveyor was to describe the area in words, and to draw a plat showing its shape and extent.

Like the maps, this interest in exact measurement was also new. Until then, what mattered was how much land would yield, not its size. When William the Conqueror instituted the great survey of England in 1086, known as the Domesday Book, his commissioners noted the dimensions of estates in units like virgates and hides, which varied according to the richness of the soil: a virgate was enough land for a single person to live on, a hide enough to support a family, and consequently the size shrank when measuring fertile land, and expanded in poor, upland territory. Other Domesday units like the acre and the carrucate were equally flexible, but so long as land was held in exchange for services, the number of people it could feed and so make available to render those services was more important than its exact area. Accurate measurement became important in 1538 because beginning in that year a gigantic swath of England—almost half a million acres—was suddenly put on sale for cash.

The greatest real-estate sale in England's history occurred after king Henry VIII dissolved a total of almost 400 monasteries, which had been acquiring land for centuries. He justified his action on the grounds that these houses of prayer had grown depraved and corrupt, but tales of drunken monks and lecherous nuns served to conceal a more mundane purpose: Henry needed money to pay for England's defenses. Upon the monasteries' dissolution, all their land, including some of the best soil in England, automatically reverted to their feudal overlord, the king. These rich acres were then sold to wealthy merchants and nobles so that a navy could be built to defend England's shores against the French.

The sale of so much land for cash was a watershed. Although changes were already under way, with feudal services often commuted for rents paid in coin, and feudal estates frequently mortgaged and sold, up to that point the fundamental value of land remained in the number of people it supported. From then on the balance shifted increasingly to a new way of thinking. Prominent among the purchasers of church property were land-hungry owners, like the duke of Northumberland, who had been enclosing common pastures, but far more common were the landlords who had done well from the rise in the market value of wool and corn, and chose to invest in monastery estates. In Norfolk, Sir Robert Southwell attracted attention because of the mighty pastures he carved out from common land for his fourteen flocks of sheep, each numbering around 1,000 animals, but others who did the same on a smaller scale almost escaped notice. In the neighboring county of Suffolk, the Winthrop family, who acquired and enclosed hundreds of acres of monastic and common land, might have remained in the background had the grandson, John, not sailed for America in 1630 as governor of the Massachusetts Bay Company.

The new owners and their surveyors realized that the monasteries' widely separated rigs and shares of common land would become more valuable once they were consolidated into fields. Their predecessors, the old abbots and priors, had understood landownership to be part of a feudal exchange of rights for services. But those who had bought their land knew that ownership depended on money passing hands, and that the old ways had to change if they were to maximize the return on their investment.

"Jesu, sir, in the name of God what mean you thus extremely to handle us poor people?" a widow demanded of John Palmer, an enclosing landlord in Sussex who, having bought the monastic estate on which she lived, had evicted her from her cottage.

"Do ye not know that the King's grace hath put down all the houses of monks, friars and nuns?" Palmer retorted. "Therefore now is the time come that we gentlemen will pull down the houses of such poor knaves as ye be."

As enclosures and rising rents forced thousands of villeins and farm laborers away from the manors that once supported them, resentment against the new owners grew rapidly. Sir Thomas More launched a particularly bitter attack on pastoralists like Southwell: "Your sheep that were wont to be so meek and tame, and so small eaters, now as I hear say, be become so great devourers and so wild, that they eat up and swallow down the very men themselves." There were popular uprisings in many parts of England, forcing Henry VIII and his successors to introduce bills in Parliament to prevent enclosure. It was a pointer to the growing influence of the landowners that few were passed.

The hidden hand in this gigantic upheaval was provided by the survey and plat that recorded the new owner's estate as his property. The emphasis in Benese's book on exact measurement reflected the change in outlook. Once land was exchanged for cash, its ability to support people became less important than how much rent it could produce. And to compare the value of rent produced by different estates, it was essential to know their exact size. The units could no longer vary; the method of surveying had to be reliable. The surveyor ceased to be a servant and became an agent of change from a system grounded in medieval practice to one that generated money.

Some at least became uneasily aware of what they were doing. In the *Surveior's Dialogue*, Norden specifically blamed the act of measuring itself for helping to destroy the old ways and held surveyors responsible as "the cause that men [lose] their Land: and sometimes they are abridged of such liberties as they have long used in Mannors: and customes are altered, broken, and sometimes perverted or taken away by your means."

What the new class of landowners required of their suveyors above all was exactness, and the sudden increase in the number of manuals in the last quarter of the century testified to the urgency of their need. In 1551 Robert Recorde wrote a book titled *Pathway to Knowledge* in praise of the accuracy that geometry offered surveyors, but warning of its potential for destruction:

> *Survayers have cause to make muche of me.*
> *And so have all Lordes that landes do possesse:*
> *But Tennauntes I feare will like me the lesse.*
> *Yet do I not wrong, but measure all truely,*
> *And yelde the full right to everye man justely.*
> *Proportion Geometricall hath no man opprest,*
> *Yf anye bee wronged, I wishe it redrest.*

It was against this background—an urgent, growing need for accurate measurement of land—that Edmund Gunter devised his chain. Born in 1581 to a Welsh family, he had been sent to Oxford University in 1599 to be educated as a Church of England priest, but discovered that numbers were more inspiring than religion. He remained at Oxford until 1615, ostensibly studying divinity, but in that time preached just one sermon and its reputation endured long after his death because, according to Oxford gossip, "it was such a lamentable one." What really interested him was the relationship of mathematics to the real world, and consequently he spent most of his time making instruments to illustrate the way in which numbers worked.

Ratios and proportions were his passion. He invented an early slide rule, known as Gunter's scale, to demonstrate proportional connections between numbers and worked out to seven places of decimals the logarithms for sine and cosine. The point at which this enthusiasm for numerical ratios touched upon concrete reality was trigonometry, which allowed mathematicians to calculate the length of two sides of a triangle when only the third side and two angles were known, but also enabled surveyors to work out the distance between two objects without having to walk between them.

It was the Dutch mathematician Frisius who handed this magic to land

measurers in 1533 by showing them how to create invisible triangles. To calculate the distance between a tower and a far-off church, for example, he
taught them first to measure exactly the distance between the tower and a
nearby object like a tree. Then, using a cross-staff or a compass, they measured the angle between imaginary lines drawn from the tree to the church
and from the tree to the tower. The next step was to repeat the operation
standing by the tower. The third angle, at the church, was calculated simply
by subtracting the sum of two measured angles from 180 degrees. The rest
was trigonometry: Knowing the angles and the length of one side, the surveyor could calculate the length of the other sides—one of which was the
distance between the tower and the church—and never have to walk a yard
of it. However, the calculations were time consuming, even with the aid of
Gunter's scale, and until logarithm tables became widely available late in the
seventeenth century, it lay beyond the capacity of most early surveyors. Nevertheless, triangulation, as this method was known, gradually became the essential method of calculating long distances, both for surveying and the
more exact science of Earth measurement or geodesy.

Since only the most basic instruments existed in Gunter's day, mathematicians and astronomers were expected to design their own, and to
demonstrate problems in geometry, Gunter himself was constantly adapting
and improving nautical instruments like the quadrant and cross-staff, which
measured vertical angles between the sun and the horizon, or horizontal angles between towers, trees, and churches. Indeed, his enthusiasm for new
gadgets cost him the best scientific job in the land. In 1620 the wealthy but
earnest Sir Henry Savile put up money to fund Oxford University's first two
science faculties, the chairs of astronomy and geometry. Gunter applied to
become professor of geometry, but Savile was famous for distrusting clever
people—"give me the plodding student," he insisted drearily—and the can-
didate's behavior annoyed him intensely. As was his habit, Gunter arrived
with his sector and quadrant, and began demonstrating how they could be
used to calculate the position of stars or the distance of churches, until Savile
could stand it no longer. "Doe you call this reading of Geometrie?" he burst
out. "This is mere showing of tricks, man!" and according to a contemporary
account, "dismisst him with scorne."

Fortunately, Gunter was supported by the earl of Bridgewater, who did like brilliance, having grown up in a house where poets like Edmund Spenser and Ben Jonson were guests and *Othello* was first performed. Since his father had inherited huge estates on the Welsh border and acquired valuable land north of London, the earl was also even richer than Savile, and it seems proba-

Gunter's quadrant.

ble that the surveyor's chain that Gunter designed in about 1607 was first used to measure the immense Bridgewater property.

Aided by aristocratic influence, Gunter was then appointed rector of the wealthy parish of St. George's, Southwark, in London and, in 1619, professor of astronomy at Gresham's College, London. However, both his congregation and students were utterly neglected in favor of his scientific instruments. Like electronic devices today, these were sold with a book of incomprehensible instructions. Gunter at least had the excuse that almost no one could understand his instructions because they were written in Latin. In 1623 the chorus of complaints persuaded him to produce a translation. "I am at the last contented that it should come forth in English," he wrote. "Not that I think it worthy either of my labour or the publique view, but to satisfy their importunity who not understanding the Latin yet were at the charge to buy the instrument."

The complete collection of Gunter's writing, issued in 1624, was titled *The description and use of the sector, the cross-staffe, and other instruments for such as are studious of mathematical practise.* By then he must have known that the last bit of the title was nonsense. The reason the book had to be in English was that his instruments were being used not by math students but by surveyors for measuring and by sailors for navigating—and unlike mathematicians, neither group could read Latin. Nevertheless, it contained so much new information on logarithms, trigonometry, and geometry that one of his contemporaries paid him this tribute: "He did open men's understandings and made young men in love with that studie [math]. Before, the mathematical sciences were lock't up in the Greeke and Latin tongues and so lay untoucht. After Mr Gunter, these sciences sprang up amain, more and more."

In this book Gunter first described the chain that was to bear his name. "For plotting of ground, I hold it fit to use a chaine of foure perches in length, divided into an hundred links." The practical appeal of a chain was that, unlike a rod, it was flexible enough to be looped over a person's shoulder, and that, being made of metal, it neither stretched nor shrank as cords always did. Other kinds of chain had been devised, but the supremacy of Gunter's came from his fascination with ratios and the relationship of one set of numbers to another. As a passionate believer in the usefulness of math, he built into his chain the most advanced numerical learning of the time until it could almost be compared to a primitive calculating machine.

Four perches measured 22 yards, a strange distance that makes sense only in the context of the traditional units used for measuring land. Like all units of land measurement, a perch, also known as a rod or a pole, originally varied according to the quality of ground—a perch of poor soil was longer than one of fertile soil—but in the course of the sixteenth century it became standardized at 16½ feet. This inconvenient length was derived from the area of agricultural land that could be worked by one person in a day—hence the variability. The area was reckoned to be 2 perches by 2 perches (33 feet by 33 feet). Thus a daywork amounted to 4 square perches. Conveniently, there were 40 dayworks in an acre, the area that could be worked by a team of oxen in a day, and 640 acres in a square mile. It was significant that all of them were multiples of 4, a number that made it simpler to calculate the area of a four-sided field.

Gunter divided the chain into 100 links, marked off into groups of 10 by brass rings. On the face of it, the dimensions make no sense: Each link is a fraction under 8 inches long; 10 links make slightly less than 6 feet, 8 inches;

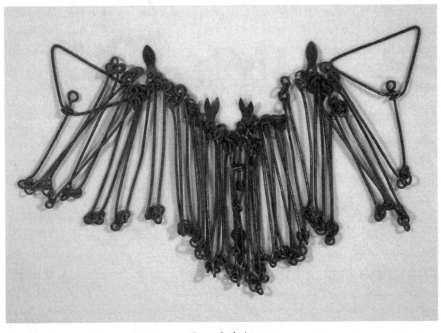

Gunter's chain.

and the full length is 66 feet. In fact, he had made a brilliant synthesis of two otherwise incompatible systems, the traditional English land measurements, based on the number 4, and the newly introduced system of decimals based on the number 10.

The Dutch engineer Simon Stevin was the first European to publish an account of decimals in 1585, and Gunter quickly grasped the concept, using them in his logarithmic tables. Decimals make arithmetical calculation simpler because ordinary numbering is based on 10: There are single digits up to 10, and an extra column is added with each multiple of it, at 100, 1,000, and so on. In effect, the decimal point is simply moved over one place. However, for a practical activity like surveying, it is far easier to halve and quarter a distance or double and redouble it. The process can be done by eye in an instant, and the result is easily checked by measurement, or if needed by calculation.

Gunter's chain allowed either method to be used. An acre measured 4,840 square yards in traditional units and 10 square chains in Gunter's system. Thus, if need be, the entire process of land measurement could be computed in decimalized chains and links, then converted to acres by dividing the result by 10. With an understandable hint of satisfaction Gunter concluded his description of its use, "then will the work be more easie in Arithmetick." It was that ease in calculating acreages, as much as its accuracy and straightforward practicality, that earned Gunter's chain its popularity among surveyors using the old 4-based system of measurements. Even the least competent could come close to the standards of exactness that were now expected of them.

It would be difficult to exaggerate the need that the growing army of surveyors had for this kind of assistance. To measure land required a combination of mathematics and technology. If the land was a square or oblong flat field, the area was simply computed by measuring the lengths of two neighboring sides, then multiplying one by the other. When the area was bounded by winding rivers and twisting coastline, and irregular features like hills and valleys had to be taken into account, the process became more complicated, and accuracy depended upon increasingly sophisticated calculation and instrumentation.

The basic technology was simple. In the field, the surveyor needed a compass to know which direction his line was running—in the seventeenth century a telescope was attached to the compass so that distant marks could be seen more distinctly, and the instrument became known as a circumferentor—and Gunter's chain to measure how far the line ran. At the end of a day working on the land, the surveyor had to transfer the distances and directions he had physically walked over into a plat or sketch map drawn to scale using dividers and a calibrated rule.

Until the sixteenth century, the math was based on the 2,000-year-old theorems of a school of Greek mathematicians, notably Euclid, who taught in the Egyptian city of Alexandria. Their study of geometry—literally "earth measurement"—was prompted by the practical difficulties faced by Egyptian surveyors reconstructing boundary markers washed away by the Nile's annual flood, and so it was appropriate that their theorems concerning the properties of triangles, squares, and circles should have been adopted by later surveyors needing to calculate the area of an unevenly shaped field.

In theory, any parcel of land, however irregular its outline, could be divided up into Euclid's regular shapes, whose areas could then simply be calculated and added together. Translating theory into practice was more complicated. "As Euclide is a Greek author," Edward Worsop observed caustically in 1582, "so is the name of his elementes Greek to a great number of measurers."

The surveyors' commonest trick when faced with an irregular shape was to add the lengths of all the sides of the area, divide the total by 4, then square the result. The answer was quick, easily worked out, and always wrong, but usually not by enough to alarm the owner. As one more scrupulous measurer observed of such shortcuts, "It is the way all Syrveyors do;—whether it originates in Idleness, inability or want of sufficient pay, it is not for me to determine." The title of Edward Worsop's volume, published in 1582, is self-descriptive: "*A discoverie of sundrie errours and faults daily committed by landemeters [surveyors] ignorant of arithmeticke and geometrie to the damage and prejudice of many of her Majestie's subjects.*"

Yet the newly landed gentry of England knew that even an imperfect survey was better than none. While they were measuring and mapping their possessions, then squeezing the highest possible rents from their tenants, the Crown ignored its own lands for seventy years after Benese's book came out. In 1603, the lord treasurer of England, Robert Cecil, at last commissioned a report on their extent and "found the King's Mannors and fairest possessions most unsurveyed and uncertain, [their area estimated] rather by report than by measure, not more known than by ancient rents; the estate granted rather by chance than upon knowledge."

Inefficiently run and casually disposed of, the royal estate, which once had produced enough to pay for much of royal government, now generated such a small income that the monarch was forced to rely on Parliament to raise taxes in order to run the kingdom. Imperceptibly, power was passing from the land-poor Crown to the land-rich gentry. And a few years later, when it was proposed to plant colonies in the newfound land of America, those who had the money to invest were the same gentry and the merchants, people who could measure their property and count its worth, not the king.

This was the power that lay in Gunter's chain—a means of making private property. So long as it was the acre that expanded or shrank while the price remained the same, no true market in land could be established. Once the earth could be measured by a unit that did not vary, supply and demand would determine the price, and it could be treated as a commodity. That was not Gunter's intention, but it was a consequence of the accuracy that was built into his means of measurement.

Precise Confusion

WHAT MADE GUNTER'S CHAIN unique among commonly used measurements was that it did not vary. There were other chains in use with different lengths, which varied according to the type of ground being measured. An early American surveying manual, for example, suggested that for open country a chain might measure 22 yards, but in woodland it should increase to 24 yards and in dense forest be as much as 32 yards. But wherever a "Gunter's chain" was specified, it meant precisely 100 links, or 22 yards, and an area measuring 10 of his chains in length by 10 of his chains in breadth would always contain 10 acres. In 1607 it would have been hard to find another measure that was so consistent.

Where modern economics would alter the price, the usual practice before the seventeenth century was to change the size of the measure. Candy bar manufacturers do the same today, because people associate a particular price with a particular candy and less resentment is caused by shrinking the bar than increasing the price. Fear of riots in the streets forced the Vatican city authorities in the fifteenth century to require bakers to sell loaves of bread at a fixed price, increasing the size when flour was cheap but reducing it when flour was dear. In thirteenth-century England, the king permitted bread sellers to change the weight of a "farthing" loaf (a farthing was a quarter of a penny) but not the price, and similar legislation on bread prices and sizes was in force in cities across Europe.

Even in the financially sophisticated Netherlands, where banks, corporations, and a stock exchange existed from the sixteenth century, it remained the custom in the textiles market to charge the same price per stone (14 pounds) for good flax from Zeeland as for the inferior kind from Brabant, but to make up the difference by using a lighter weight for the stone. The Dutch milling trade followed a similar pattern: selling flour in larger measures to wholesalers than to retailers, but charging the same price per measure to both. At the other end of Europe, in the port of Riga—the center of trade in the Baltic and a major partner in the trading association known as the Hanseatic League—the Latvians quoted a single price on the salt fish they sold to Russian merchants, Ukrainian nobility, and their Hanseatic partners, but measured them out in different scales and containers according to the importance of the customer. In France, traveling salesmen charged a standard price per bushel of flour, but in remote villages used a smaller container than in those nearer to town, and explained that "the differences in measures are such as to defray the costs of transport."

Just as an acre of rough pasture was larger than an acre of meadow, which could produce hay, so a bushel of oats, which would only make porridge, held more than a bushel of wheat, which could be made into bread, and weak beer suitable only for quenching the thirst was gauged by a large gallon, whereas wine that intoxicated the mind was sparingly measured in a small gallon. The principle applied equally to textiles, wood, and other commodities. An ell of coarse cloth was longer than an ell of fine material, a cord of green firewood bulkier than one of dry. Each varied not only according to place but according to the benefit it yielded to the buyer. In short, what was being measured was not a quantity but its local, subjective, human value.

The measures themselves were derived from the shape of the human body itself and from human activity . Consequently, even a basic measure of length like the inch, which had been inherited from the Romans, could not be exactly defined. The inch was reckoned to be the width of a finger (usually 16 to the foot) or a thumb (usually 12 to the foot), and in the first century

B.C.E., the Roman Vitruvius Pollio gave the classic definition of its relationship to other measures: "Four fingers make one palm, and four palms make one foot; six palms make one cubit; four cubits make once a man's height." But since fingers and thumbs came in different sizes, so did the inch and the other units. The size of a bushel originated with the amount of seed required to sow an acre of ground—it thus varied as much as the size of the acre— while the ell, used for cloth, was either the width of the loom or the distance from head to wrist. (The easiest way to measure woven material is to hold it out with an outstretched arm from the chin.) Equally organic, and still less exact, were units like the English bowshot (the distance an arrow would fly), the French *houpée* (how far a shout would carry), and, among the Plains Indians of America, a horse-belly view (the farthest a person could see over the prairies when squatting beneath a mustang, approximately 2 miles).

Inevitably, variable measures offered irresistible opportunities for cheating. Since the first written records, and probably earlier than that, there have been denunciations of those who used false measures. The commonest deceit was simply to use two sets of weights and containers, the large one for buying, the small for selling, and throughout history, sacred and secular authorities have thundered against the practice. "Thou shalt not have in thy bag diverse weights, a great and a small," runs the Jewish law in the book of Deuteronomy. "But thou shalt have a perfect and just weight, a perfect and just measure shalt thou have, that thy days may be lengthened in the land which the Lord thy God giveth thee." The Koran inveighs in similar fashion: "Woe to those who stint the measure. Who when they take by measure from others, exact the full, but diminish when they measure to others, or weigh to them."

In 813 Charlemagne, newly crowned as emperor of most of western Europe by the pope, issued a famous edict, which began, "Volumus ut pondera vel mensurae ubique aequalia sint et iusta." (We desire that weights and measures should be equal and just everywhere.) For nearly 1,000 years thereafter, the goal of almost every French king could be expressed euphonically as "un roi, une foi, un poids" (one king, one faith, one weight). In 1543 Francis I asserted bluntly that "the supreme authority of the King incorporates the right to standardize all measures throughout his kingdom." But in 1790, according to an authoritative estimate, France possessed 13 separate lengths for a *pied* (foot), 18 for the *aune* (ell), and 24 for the *boisseau* (bushel), and since the seventy-four parishes of Angoulême near Bordeaux boasted more than 100 sizes of *boisseau* between them, with one parish alone offering 4 separate varieties, this was something of an underestimate.

The existence of variable measures, with all their opportunities for fraud, stifled any wider trade because only those actually watching the wheat bushel being filled or the linen being stretched out knew the quantities involved. The success of the great medieval trade fairs at Troyes in Champagne was largely due to the town's capacity to impose its own measures on traders, but it was the Dutch, the first great exporting nation in Europe, who understood the advantage of accuracy better than anyone. According to Josiah Child, writing in 1668, the booming trade that produced their astonishing prosperity in the seventeenth century was largely based on "their exact marking of all their Native Commodities." Their major export was fish—herring and cod packed into huge barrels called hogsheads, which were supposed to contain 64 gallons. Child noted that the Dutch measures were so reliable "that the repute of their said Commodities abroad continues always good, and the Buyers will accept of them by the marks, without opening; whereas the Fish which our English make in New-found-Land and New-England, and Herrings at Yarmouth . . . often prove false and deceitfully made, seldom containing the quantity for which the Hogsheads are marked."

The obvious advantages of exact measuring should have made it easy to impose uniform, reliable measures, but they remained local and changeable for a reason that is fundamental to the history of weights and measures. However much emperors and rulers might legislate for uniformity, the actual scales, yardsticks, and containers were held by city authorities and landed nobles. They were a source of such profit that no one gave them up unless forced to. As a result, control of weights and measures has always offered a good indication of who exercised day-to-day authority. Just as the power of the Dutch trading guilds was demonstrated by their ability to establish uniform measures for exports, so the bewildering variety in France was testimony to the feudal power of the seigneurs, who retained the right to regulate local weights and measures. In England the emergence of Gunter's chain as the one measure to determine the dimensions of landed property was due not just to its practicality but to the power of a particular class of people.

*O*N THE FACE OF IT, England was in much the same position as France. In 960 King Edgar declared that "the measure of Winchester [England's capital] shall be the standard" for the whole kingdom, but the number of monarchs after him who also demanded uniformity—the call for "one weight and one measure" appears identically in Richard the Lion-Heart's decree of 1189 and twenty-six years later in the Magna Carta—suggests that they were no more effective than their French counterparts.

The most important of these medieval laws, enacted by Henry III in 1266, introduced the sterling system linking weights to coinage, so that there were 240 pennyweights to the pound. This ratio persisted in the currency for more than 700 years (until 1972) in Britain, and in North America until it was superseded by the dollar. Nevertheless, most of Henry's successors still found it necessary to pass laws against "false and deceitful measures," and in 1496 Henry VII took the curiously modern step of dumping the sterling pound in favor of a European unit, the troy pound. Yet in 1588, less than a century later, his granddaughter, Elizabeth I, had to introduce still more legislation, which she explained was "called forth by the uncertainty of the weights then in use, to the great slander of the realm and decency of many, both buyers and sellers."

It is against this background of incessant variation, deceit, and falsehood in weights and measures that the precision of Gunter's chain needs to be set. More than any of her predecessors, Elizabeth was responsive to the power of the House of Commons and the people represented there. The contrast between the way she legislated for weight and for length and area was significant.

Where weight was concerned, she found it necessary to add to the troy system the heavier avoirdupois, which literally means "having weight," range, which went from ounces through pounds and stones to hundredweights and tons. Troy was ideal for measuring small items like gold and silver, but the main English export was wool, which was traded in elephantine quantities and in the Flemish markets was always weighed in avoirdupois. It was a concession to variability to use a goldsmith's troy weights for light objects and a wool merchant's avoirdupois for heavy ones, and a similar surrender appeared in her decision to legislate for a large gallon for measuring beer and a small gallon for wine. Adding to the confusion, careless wording of the specifications for containers resulted in four sizes of bushel being legalized for measuring grains and flour.

By contrast, the specifications for length were both simple and accurate. The critical decision was to define for the first time in 1595 the length of a mile at 1,760 yards instead of the outdated, Roman distance of 5,000 feet or 1,667 yards. The new length was a concession to surveying needs. It represented 8 furlongs of 220 yards each, and the easiest way to measure out an acre was in the form of an oblong, 220 yards long by 22 yards broad, or, in a surveyor's terminology, 40 perches by 4 perches.

Four was the surveyor's base figure, and the fourness of Elizabeth's measures was striking. "Foure graines of barley make a finger," stated a rule of 1566, "foure fingers a hande; foure handes a foote." This produced a foot of 16 small inches or fingers, but surveyors were not interested in such small quantities, and the more convenient foot of 12 larger inches gradually superseded it. The critical unit was the yard, which went to make up the perch, the furlong, and the mile, and here the precision was remarkable. In 1601 a brass yardstick was constructed as a standard for the country as a whole. Exactness was what the market required, and when Elizabeth's yard was measured in 1797 against the inches, feet, and yards used by eighteenth-century scientists, it was found to be precisely 36.015 inches long.

Elizabeth had the energy and administrative skill of a great ruler. She not only ordered new standards to be made for these weights and measures but sent copies to fifty-eight market towns with instructions that a description of them was to be pinned up in every church and read during the service twice a year for the next four years. For good and ill, it is to her that the credit must go for creating a system of weights and measures that was to persist for almost 400 years, eventually covering all of Britain and almost one quarter of the globe.

All in all, it was a measuring age. Accurate measurement was becoming vital to the navigation of England's mariners, who used Gunter's cross-staff and quadrant to find latitude in the trackless ocean, and most notably to Sir Francis Drake in circumnavigating the globe between 1577 and 1580. It was critical to the founding father of the scientific method, Francis Bacon, who advocated measurement and experimentation as the basis of science. When Elizabethans met, they took one another's measure, they danced tightly paced measures like the galliard and volta, and they measured their poetry to the short-long rhythm of iambic pentameters. "Marry, if you would put me to verses, or to dance, for your sake, Kate, why you undid me," exclaims rough King Harry wooing his French princess with puns in Shakespeare's *Henry V.* "For the one I have neither words nor measure, and for the other I have no strength in measure—yet a reasonable measure in strength."

It was a joke tailored to that particular audience. Without measure, music was noise, poetry babble, and the land wilderness, and none knew it better than the enclosing, acquisitive gentry, the generation whose parents and grandparents first bought their land from Henry VIII, who stamped the Elizabethan age with their energy and imagination, and for whose benefit the legislation on measures was passed.

John Winthrop was just such a man. His family had acquired their 500-acre estate of Groton Manor from Henry VIII, and he himself was a vigorous encloser and improver of the land. It was as much the downturn in rents and farm prices as his Puritan ideals that persuaded him in 1630 to volunteer to take charge of the colony that the Massachusetts Bay Company proposed to create in Boston. Authoritarian, clear-sighted, and charismatic, he was the colony's first governor and imbued it not only with his ideals of communal responsibility and individual conscience but with his attitude toward property.

Although the royal patent gave the colonists the right to settle in New England, there were those, notably Roger Williams, founder of the Rhode Island colony, who believed that the land rightly belonged to the native inhab-

John Winthrop

itants and should first be bought from them. Winthrop summarily disposed of that view with an argument grounded in his own upbringing. "As for the Natiues in new England," he wrote, "they inclose no Land, neither haue any setled habytation, nor any tame Cattle to proue the Land by."

Since the Native Americans had nothing to show that they owned the land, the new Americans could take it freely, and New England, like the Old, would belong to those who could measure it and enclose it.

THREE

Who Owned America?

THEIR FIRST GLIMPSE of the land across the Atlantic provoked in the early colonists a remarkably similar mixture of emotions. Most reacted as John Bereton did when he landed in Massachusetts in 1602; as he later recalled, "We stood a while like men ravished at the beautie and delicacie of this sweet soile." But throughout the seventeenth and eighteenth centuries land was the prime source of wealth—the rewards of trade were unreliable and restricted to the minority who lived in cities, while industrial wealth was so rare as to be freakish. Thus, soon after that first sense of almost spiritual awe at the beauty of the country came a sharp consideration of the profit it represented.

"The mildnesse of the aire, the fertilitie of the soile, and the situation of the rivers," wrote Captain John Smith in his famous account of Virginia in 1624, "are so propitious to the nature and use of man as no place is more convenient for pleasure, profit, and mans sustenance."

Even a Jesuit priest like Andrew White, confronted by Maryland's untouched country, confessed to the same greedy delight. "I will end therefore with the soyle, which is excellent," he wrote in 1634, "so that we cannot sett downe a foot, but tread on Strawberries, raspires, fallen mulberrie vines, acchorns, walnutts, saxafras etc: and those in the wildest woods. . . . It abounds with delicate springs which are our best drinke. Birds diversely feathered there are infinite, as eagles, swans, hernes [herons], geese, bitterns, duckes, and the like, by which will appeare, the place abounds not alone with profit, but also with pleasure." And he ends with a heartfelt "*Laus Deo*—Praise be to God."

The potential for profit made it vital to determine who owned this rich territory. By Winthrop's definition, the native inhabitants had no real claim because they had not enclosed it. So far as he and colonists like him were concerned, legal ownership of the land they settled belonged ultimately to King James I as their feudal overlord, and the king had used that power to transfer the right of possession to the companies that had invested in the colonies on the other side of the Atlantic. But ownership of land is never simple. It includes rights not just to the soil but to the metals below, the vegetation above, the sunlight, and the air; to the use, development, access, and enjoyment of the land; and to much more that, for a fee, any lawyer will reveal. Since any or all of these may be bought, rented, leased, and distributed in different ways, landed property is usually described as a bundle of legal rights that can be split up and dealt with separately.

Consequently, much of the 1629 charter creating the Massachusetts Bay Company was made up of lists of different types of land, forms of ownership, and the way they were to be transferred; the king promised to "give, graunt, bargaine, sell, alien, enfeoffe, allot, assigne and confirme" to the company all the "Landes and Groundes, Place and Places, Soyles, Woods and Wood Groundes, Havens, Portes, Rivers, Waters, Mynes, Mineralls," et cetera. Nevertheless, ultimately Massachusetts remained the king's, and the company would hold it of him "in free and common soccage"—which meant that having given, granted, and all the rest, the Crown retained an overriding, feudal competence over·that part of America.

The division of responsibility was shown by who measured what. The king defined the outer limits of British America by map references given in the royal charters, but it was the company that appointed the surveyors to lay out the boundaries of the colonies it had been granted. Thus King James I's 1609 charter to the two companies who had put up the money for the Virginia plantation specified that the London company was to plant its colony "in some fit and convenient Place, between four and thirty and one and forty Degrees of the said Latitude," and the west of England company based in

Plymouth was allocated "some fit and convenient Place, between eight and thirty Degrees and five and forty Degrees."

The four-degree overlap was reduced in 1620 when a new charter gave the Plymouth company all the land, to be known as New England, "from Fourty Degrees of Northerly Latitude, from the Equnoctiall Line, to Fourty-eight Degrees of the said Northerly Latitude." Similar charters delineated the geographic limits of all the Atlantic colonies from Nova Scotia to Georgia, often with a final phrase extending their width "to the South Sea," in other words, to the Pacific Ocean.* A few, like Maryland and Pennsylvania, had western boundaries fixed in lines or meridians of longitude.

It was the responsibility of the companies and later of the royal and aristocratic proprietors named in the charters to have those lines of longitude and latitude, so easily described in the Privy Chamber in Whitehall, marked out on the ground. The task provoked a sustained wail of complaint from the surveyors who ran the boundaries between the colonies. It was one thing to follow the lay of the land, as the settlers did, zigzagging up from the coast, following rivers and valleys into the foothills of the Blue Ridge Mountains or the Alleghenies, quite another to run a straight line up the hills, through the swamps, and into the unending forest until they emerged into the savannas of the piedmont. Nevertheless, to establish their rights, the owners had to have the boundaries of their colonies and plantations run westward from the coast.

The most formidable obstacle was the Great Dismal Swamp, a 900-square-mile expanse of stagnant water, dense bamboo groves, and crowded, vine-choked trees, lying on the border between Virginia and Carolina. In his account of marking out that border in 1728, *The History of the Dividing Line*, William Byrd II, one of the boundary commissioners, described the surveyors' approach to the swamp. "The Reeds which grew about 12 feet high, were so thick, & so interlaced with Bamboe-Briars, that our Pioneers were forc't to

---

*The colonies were created by royal charter or grant in the following order: 1606, Virginia; 1610, Newfoundland; 1620, New England; 1621, Nova Scotia; 1629, Massachusetts Bay; 1632, Maryland; 1635, New Hampshire; 1662, Connecticut; 1663, Carolina; 1663, Rhode Island; 1664, New Jersey; 1674, New York; 1681, Pennsylvania; 1701, Delaware; 1732, Georgia.

open a Passage. The Ground, if I may properly call it so, was so Spungy, that the Prints of our Feet were instantly fill'd with Water. But the greatest Grievance was from large Cypresses, which the Wind had blown down and heap'd upon one another. On the Limbs of most of them grew Sharp Snags, Pointing every way like so many Pikes, that requir'd much Pains and Caution to avoid."

Undeterred, the lead surveyor, William Mayo, pushed through the reeds and disappeared from sight. On the far side of the swamp, Byrd and the other commissioners waited anxiously. After a week, they started to fire off muskets to attract the surveyors' attention, but with no success until on the ninth day the mud-stained party at last emerged, having run the boundary through 15 miles of swamp.

Similar hardships faced the surveyors running the boundary between North and South Carolina in the 1730s, when the state was split into two. Complaining of the "Extraordinary fatigue [of] Running the said Line most of that time thro' Desart and uninhabited woods," and over rivers and marshland that were breeding grounds for snakes and clouds of vicious mosquitoes, the first team gave up after a few miles. Thirty years later, James Cook from North Carolina took on the task but, distracted by "the rains, the hot weather and the insects"—or so he claimed—ran the line 11 miles south of the thirty-fifth degree, which was the intended boundary, and thus took 660 square miles from South Carolina for the benefit of his own state. It was a potent illustration of the surveyors' power in determining boundaries, and it was certainly not the only mistake.

In 1688 John Love published the first surveying manual designed specifically for American conditions, *Geodaesia or the art of surveying and measuring of land made easie,* with the subtitle *How to lay-out new lands in America, or elsewhere.* As he explained, having worked in Carolina for some years, he had discovered that many surveyors there did not even know how to lay out a property "when a certain quantity of Acres has been given to be laid out five or six times as broad as long. This I know is to be laught at by a Mathematician; yet to such as have no more of this Learning than to know how to Measure a Field, it seems a Difficult Question."

His book highlighted the particular problems faced by American surveyors of running a line through thick woodland or across uncharted country compared with the English experience of measuring long-established, often well-cultivated landholdings. He insisted on his readers learning some very basic math, including square roots, and showed them how to use instruments like the circumferentor or compass, the semicircle for measuring the vertical angles of hills, and the plane-table for horizontal angles. But in case this was too demanding, he also demonstrated how to survey the land using only Gunter's chain because, in the words of one enthusiast, "he that can but add, subtract, multiply and divide is sufficently qualifyd with Arithmetick to practise this Method of Surveying."

What this implicitly recognized was the existence of different grades of surveyor, and their equipment provided the best guide. At the most basic level, every measurer carried a 16½-foot rod, or Edmund Gunter's invaluable chain, but a good professional would also use a circumferentor, which by the eighteenth century had developed into a theodolite or transit with crosshairs in the lens of the telescope, and built-in compass and plumb line. For the experts, a quadrant or sextant for making sun sights to check their position would also be included, but when Charles Mason and Jeremiah Dixon were hired in 1763 by the proprietors of Pennsylvania and Maryland to sort out the disputed boundary between their two provinces, their equipment included a zenith sector built by John Bird, London's foremost instrument maker. This was a telescope almost 6 feet long, exactly calibrated and pointing vertically, beneath which they lay flat on their backs to take sightings on particular stars as they passed precisely overhead. Star charts showing the positions of those stars at different dates and latitudes then enabled them to calculate their latitude with great precision.

It cost the Calverts of Maryland and the Penns of Pennsylvania the immense sum of £3,500 to have the 244 miles of what became known as the Mason-Dixon line surveyed with such accuracy. To them it was money well spent, for in a new country there was no other way of establishing ownership. Without a survey, anyone could claim the land. Nowhere was that more apparent than in Pennsylvania, which was founded not just on Quaker toleration but on the surveyor's chain.

William Penn's vision for his "holy experiment" depended on measuring out the land, beginning, as he promised the original settlers in 1681, with the capital: "A certain quantity of land, or ground plat, shall be laid out, for a large town or city, in the most convenient place, upon the river, for health and navigation." The original shape, drawn by his surveyor, Thomas Holme, can still be found in Philadelphia's gridiron center, hemmed in between the Schuylkill and Delaware Rivers, with Penn Square at its heart, measuring 10 chains by 10 chains as specified by the plat. Similar, smaller settlements were to be built across the colony, and early maps show that close to his capital and along the coast, villages were laid out in the rectangular shape that Penn wanted, with the houses lying near to each other "so that the neighbors may hold one another in a Christlike manner and praise God together."

The royal charter granted him in 1681 gave him the legal right to enforce these conditions on the settlers, and the fact that they paid him an annual quitrent, which took the place of such feudal duties as road mending and bearing arms, showed the extent of his powers.

In the Carolinas, the proprietors used their power to decree in their 1665 constitution that all the lowland area, the tidewater, should be presurveyed by a surveyor-general and divided into squares and rectangles "by lines running East and West, North and South." From these blocks they proposed to build an American aristocracy, with ordinary immigrants receiving a grant of 100 acres and paying a feudal quitrent on them, and above them proprietors, lords of the manor, and lesser nobles whose rank depended on the size of their landholding. To ensure compliance, the proprietors instituted a complex system that required the settler to obtain a land warrant from the governor, followed by a survey from the surveyor-general, before the land could be allocated and the claim registered.

In Georgia, the proprietors, and particularly the idealistic James Oglethorpe, also intended to survey its territory before distributing it in order to create a slave-free, teetotal society of smallholders and farmers living close to each other in rectangular properties. In the beautifully proportioned squares and gardens of its capital, Savannah, can be seen all that remains of the plan.

Had the companies and proprietors succeeded in maintaining control over the land they held from the king, the history of North America might have remained colonial. But the idea of property that the colonists carried with them created its own revolutionary current.

The first years after the Pilgrim fathers landed in Plymouth in the bitter winter of 1620 indicated the direction that history would take. Under the terms of their agreement with their financial backers, they were to work the land in common, sharing the proceeds with the investors in England. The goal of communal landownership should have been particularly attractive, for they arrived with a close sense of unity arising from the shared desire for religious freedom. Yet in the first years, when they attempted to pool their resources and farm collectively, with young men assigned to work for those who had families, the fields were neglected and they almost starved. In desperation, Governor William Bradford responded to demands that the land be divided up. "And so," he noted in his history of the Plymouth Colony, "[I] assigned to every family a parcel of land according to the proportion of their number. . . . This had very good success for it made all hands very industrious."

The dramatic increase in yields soon ensured their food supplies, but the change came at a cost. "And no man now thought he could live except he had catle and a great deale of ground to keep them all," Bradford observed sadly, "all striving to increase their stocks. By which means they were scatered all over the bay quickly and the towne in which they lived compactly till now was left very thinne." Religious freedom might have been their prime reason for sailing to America, but once they were there, the desire to own land was too powerful to be ignored. As Richard Winslow put it in his 1624 pamphlet *Good Newes from New England,* "Religion and profit jump together."

In 1691 the thinly populated colony was absorbed into the wealthier Massachusetts Bay Colony. But it too had changed since Winthrop had founded it as the shining light of Puritanism, "the city upon a hill." By then, the Puritan preacher Increase Mather was lamenting that the grandchildren of the original settlers had grown insatiable for land. "How many men have since coveted after the earth," he thundered, "that many hundreds nay thousands of acres have been engrossed by one man, and they that profess themselves Christians have forsaken churches and ordinance, and for land and elbow-room enough in the world?"

In Virginia, the first Jamestown colonists never had that religious sense of cohesion. They saved themselves from starvation only by raiding the farms of Powhatan Indians, but fifteen years after the first settlers arrived in 1609, massacre and disease had killed 6,000 out of 7,300 migrants from England. The colony's only source of income came from the sale of tobacco, and that was not enough to prevent the Virginia company from going bankrupt in 1624. What kept the colony alive was a decision in 1618 by one of its shareholders, Sir Edwin Sandys, to attract immigrants by offering a "headright" of 50 acres of good Virginia soil to anyone who crossed the ocean at his own expense, and as much again for every adult he brought with him. The lure of free land brought a stream of would-be settlers, most of whom died, but by the 1630s the flow of migrants outstripped the death rate from fever, and soon land was being bought and sold at five shillings (about $1.25) for 50 acres.

*B*Y THE END OF THE seventeenth century, the shape of British America was long and thin, stretching from thirty-one degrees north to forty-nine degrees north, a distance of over 2,000 miles but, so far as measured, settled land was concerned, rarely more than 200 miles deep, a sort of northern Chile. It was in the first years of the eighteenth century that siren voices from beyond the swamps and pine barrens began to tell of the irresistibly fertile ground to be had in the piedmont or foothills of the Appalachian Mountains. "The best, richest, and most healthy part of your Country is yet to be inhabited," wrote Francis Makemie in *Plain and Friendly Persuasion* in 1705, "above the falls of every River, to the Mountains."

The first pressure was felt in Pennsylvania, where thousands of Moravians arrived from Germany's Rhineland in the early eighteenth century and chose to settle in the Lancaster Valley, set among rolling hills beyond the reach of Penn's surveyors. They were followed by Scotch-Irish Presbyterians, who resolutely ignored every attempt by the Penn family to restrict settlement to approved areas. The Crocketts, Paxtons, and Houstons had left Ulster to find religious freedom in Pennsylvania and swarmed into the area beyond Lancaster, squatting on the vacant land, with the bold argument that "it was against the laws of God and nature that so much land should be idle

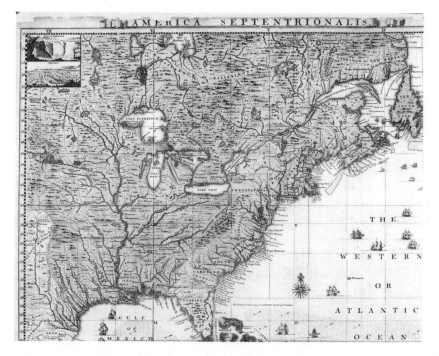

The British colonies in America, 1733.

while so many Christians wanted it to work on and to raise their bread." By 1726 the colony had an estimated 100,000 squatters, and two-thirds of it was occupied illegally.

To the north and south, similar motives were driving new settlers and the children of earlier generations to move outside the patterns of habitation laid out by proprietors and royal governors. In the Virginian piedmont, William Byrd watched crowds of Scotch-Irish squatters taking whatever land suited them and was reminded of "the Goths and Vandals of old." In Massachusetts, settlers moved out into the hilly Berkshires, and in New York up the Mohawk Valley, constrained only by fear of French and Indian attack. In an attempt to retain control, the Massachusetts government established a string of new townships on the New Hampshire border and in the Berkshires during the 1730s. Elsewhere, proprietors in Maryland and Pennsylvania, great landowners in New York and northern Virginia, either offered leases to squatters on their land or tried to drive them out.

The mighty five-million-acre estate of Lord Fairfax in northern Virginia, almost a state within a state, was infested with illegal settlers. The family responded by establishing its exact extent so that squatters already on the land could either be persuaded to take on a ninety-nine-year lease at around five dollars a year for 100 acres, or be forced out. A survey made the land profitable—the lack of one left it vulnerable to squatters, or, as exasperated landowners preferred to call them, *banditti*—and the task was so important that no matter what the terrain, the Fairfaxes were determined the boundaries should be run.

In 1746 Lord Fairfax employed Thomas Lewis and Peter Jefferson, father of Thomas, to mark off the southern boundary, stretching into the Blue Ridge Mountains. In his diary, Lewis wrote of their extraordinary hardships. "[Descending in the dark from a mountain,] we fell into a place that had precipices on either [side], very narrow, full of ledges and brush, and exceedingly rocky. A very great descent. We all like to been killed with repeated falls, and our horses were in a miserable condition. The loose rocks were so [dangerous] as to prove fatal. We at length got to the bottom, not much better, there being a large water course with banks extremely steep that obliged us to cross at places almost [vertical]. After great despair, we at length got to camp about 10 o'clock, hardly anyone without broken [bones] or other misfortune."

Ten days later the line cut across a swamp "full of rocks and cavities covered over with a kind of moss [to] considerable depth. The laurel and ivy [were] so woven together that without cutting it [was] impossible to force through." And when they finally got out of the swamp, Lewis wrote in heartfelt relief, "never was any poor creatures in such a condition as we! Nor ever was a criminal more glad of having escaped from prison as we were to Get Rid of those Accursed Laurels! From the Beginning of Time, when we entered this swamp, I did not see a [dry] place big enough for a man to lie nor a horse to stand."

Surveyors who were prepared to undertake such hardships, and could produce a reliable plat, were so valuable to a landowner they could command enormous fees, and the demand was such that by the mid–eighteenth century a good practitioner could command an income that matched a lawyer's. Even at the age of seventeen, George Washington, who began his career as a surveyor by working for the Fairfaxes, was able to boast of his earnings to a friend. "A doubloon is my constant gain every day that will permit my going out," he wrote, "and sometimes six pistoles." And since a doubloon was worth about $15, and "six pistoles" around $22.50, a good week might bring in around $100. Even as president, he barely earned more.

As the pressure from illegal settlers increased, royal governors in Virginia and the Carolinas invented schemes to make the squatters legal by offering to lay out new townships on the frontier where the land would be free land. None of it worked. The prospect of so much property was irresistible. In Virginia, Governor Alexander Spotswood himself succumbed to the land rush and claimed 85,000 acres of the newly opened uplands for himself, and in the Carolinas the system of land allocation was overwhelmed by the demand for surveys. Amateur surveyors were hired to help. Wildly unrealistic plats were registered. No one minded. Within two years, warrants were issued for about 600,000 acres, and 19,000 of them went to the governor, while the assembly members voted themselves 6,000 acres apiece.

Between 1731 and 1738 approximately one million acres were registered in Carolina, and when the surveyor-general, Benjamin Whitaker, complained in 1732 that "the law enables any common surveyor to perpetuate frauds for his employers through not having to turn his survey into any office," the outraged assembly sent him to prison for contempt. By then, the proprietors had lost all control and in despair turned the colony over to the Crown.

In Georgia, the demand for land blew apart the idealistic vision that Oglethorpe and his fellow proprietors entertained. Tempted by the prospect of unclaimed, fertile territory, Carolinian planters moved south across the border in increasing numbers and were soon imitated by tidewater Georgians moving inland. The squatters claimed vast estates beyond the Savannah River, outside the squares surveyed by Georgia's founders. In 1751 the colony's proprietors followed the example of Carolina and returned the colony to royal control.

Eventually, all settlers, legal or illegal, needed to register their claims in order to establish ownership, and this required the bare minimum of a survey, a plat, and entry on the colony's land registry. But the method of surveying opened up a critical divide between North and South.

In Virginia, settlers were always allowed to choose their own parcel of land first, and then to have it surveyed and registered. Both the 1,000-acre tobacco plantations awarded to shareholders and the 50- or 100-acre "headright" farms granted to each colonist who paid his own passage were established in the same way. The planter selected the area—usually the bottomland along a navigable river with some nearby woodland to provide building material—then crudely gauged the acreage he needed. Each property was reckoned to run back for a mile from the riverbank. Using Gunter's chain, the settlers simply measured out a length of 6 chains and 25 links. This produced a seemingly awkward distance of 137 yards, 1 foot, and 6 inches, but multiplied by the 1,760-yard depth of the farm, it gave a total of 242,000 square yards, or precisely 50 acres. A 100-acre farm was 12 chains and 50 links broad, and a shareholder entitled to 1,000 acres measured out 133 chains. This was frontier math, and it became second nature to anyone who wanted to own land. Once the claim had been chosen, it would be surveyed by the county surveyor and the plat entered on the land registry.

Although the first farms and plantations were more or less square, later arrivals fitting in around them produced crazy patterns of settlement that challenged the limited skills of surveyors. To define the boundary of their property, the owners blazed trees or scratched boulders or raised mounds, and described their holdings in terms of these markers. This was the old English practice of "metes and bounds," and almost inevitably, it was this method that took over when the proprietors' plans to presurvey the land broke down in Carolina and Georgia. In the well-settled countryside of England, "metes and bounds" presented few problems, but in the trackless American woods, the combination of inaccurate survey plats and boundary markers that might easily burn down or wash away was to create a snakepit of disputes. Its impact on southern landholding was so profound that the effects can still be seen almost 300 years later.

In New England, where the climate and soil were harder and the first colonists arrived as united, religious groups, a different system evolved, which placed the emphasis on communities rather than individuals. Land was allocated to congregations or churches by the General Court, which appointed a surveyor to lay out the township—usually in rectangular blocks, 6 or 10 miles square—before the new settlers moved in. In 1636, for example, William Pynchon, Jeheu Burr, and half a dozen others were given permission to create a settlement for forty to fifty families at Agawam, just west of the Connecticut River. Each family was to have enough property for a house and "in-lot" of farmland near the center of the settlement, and an "out-lot" beyond consisting of a "hassocky marsh" and upland hills for grazing. The precise width of each house lot was laid down: "Northward lys the lott of Thomas Woodford beinge twelve [rods] broade and all the marish before it to ye uplande. Next the lott of Thomas Woodford lys the lott of Thomas Ufford beinge fourteene rod broade and all the marish before it to ye uplande. Next the lott of Thomas Ufford lys the lott of Henry Smith beinge twenty rod in bredth and all the marish before it."

No gaps were left between one individual holding and the next, and one township and the next. The northern settlers might not have been able to choose the precise parcel they wanted, but they enjoyed one advantage over the southern planters. In the South, the last remnant of feudalism required landowners to pay the proprietor or colonial government an annual quitrent of up to two shillings (about fifty cents) an acre, to be quit of the obligations and services they would otherwise owe as vassals. Failure to comply would result in a notice ordering the owner "to pay [his] arrears of Quit-Rents and Reliefs and to make [his] Oath of Fealty" or be fined. In New England, the complication of levying it through the church or town soon led to the quitrent being abandoned, which meant that freeholders in a New England town effectively owned their farms in fee simple—free of all feudal dues and obligations. They had other social duties—to pay the minister's salary and attend the church or meetinghouse—but their land was undeniably their own.

In the eighteenth century, it was the lure of owning land that, above all, drew in the growing flood of settlers. If they could not pay their own passage, they came as indentured servants, willing to act as near-slaves for a number of years for the chance of acquiring property. Even in the seventeenth century, many English colonists had been villeins and laborers who were forced to emigrate when the enclosures deprived them of the common land and common grazing they depended on to keep their families alive. That process accelerated in the eighteenth century, and when enclosure reached Scotland, improving landlords in the lowlands and clan chiefs in the highlands took as their own the land that their tenants and clansmen once held in common, providing more raw material for the colonies. There was a certain irony that these should now be among the hungriest of all America's property owners, relying on the surveys and chains that originally drove them off their homeland, but none knew better than the dispossessed the urge to own a farm that could not be taken away.

By 1751 the population of the Atlantic colonies had risen to more than one million, far outstripping that of New France and outnumbering the Spanish-descended inhabitants in New Spain who had been there for more than 200 years. The explosion in numbers was fueled partly by the need for large families to work the land, but still more by the lure of property. No one seemed to find it strange, yet until the middle of the eighteenth century this restless hunger to own land made the British colonies unique in North America.

In Mexico and up the Pacific coast, the Spanish acquired land as part of a general pattern of royally sponsored exploration and settlement by the king's representatives. A Spanish civilization was created in Mexico with a university, a bishop, and a capital housing some 15,000 Spaniards before the first English colonist landed in Massachusetts. The Law of the Indies, enacted in 1573, specified in detail how the colonial government was to lay out towns and settlements. The sites were to be surveyed; then religious missions would be created to convert the natives, military presidios to defend the colonies, and civilian pueblos where colonists and colonized could live. It was an empire created from above, belonging to the king, and administered as a royal dominion. Individual landholdings were so rare that during the half-century when Spain ruled California, from 1769 to 1821, fewer than thirty families were permitted to acquire their own ranchos, or estates.

In the 150 years following Samuel Champlain's establishment of an armed post at Quebec in 1608, New France was ruled almost as rigidly. A string of trading ports was established along the St. Lawrence River, as far as the Great Lakes, and down the Mississippi to the Gulf of Mexico. Cities like

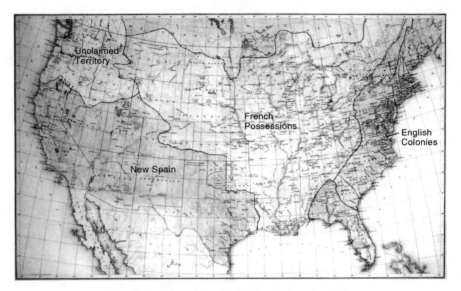

British, French, and Spanish holdings in America, 1755.

Montreal and New Orleans were founded, and farms were cultivated in Canada and in the Mississippi Delta. Nevertheless, French America was administered feudally. The Crown owned the land and chose who could settle there; Protestants, for example, were banned. It created monopolies to exploit the fur and timber. The habitant who actually worked the soil never had clear rights to it. What he owned was the use of the land and the improvements he made to it, but he held the land from a seigneur in return for dues and rents, and the seigneur held the land from the Crown. French traders and trappers knew the country intimately—they supplied British mapmakers with much of their geographic information—yet by the middle of the eighteenth century barely 40,000 had acquired land outside the main cities.

The land hunger of the British colonists seemed most extreme when set alongside the attitude of the Native Americans. From the farming Powhatan in Virginia to the Iroquois in New York and the Six Nations in the Appalachians who were primarily hunters, they shared a pervasive understanding that a particular place belonged to a particular people only to the extent that the people belonged to the place. Rights over land were gained only by occupation, long usage, or family burial, and they were communal, not individual, rights. "What is this you call property?" Massasoit, a leader of the Wampanaog, asked the Plymouth colonists whom he had befriended in the 1620s. "It cannot be the earth, for the land is our mother, nourishing all her children, beasts, birds, fish, and all men. The woods, the streams, everything on it belongs to everybody and is for the use of all. How can one man say it belongs only to him?"

Yet the British colonists bought and sold land as though they owned it outright—in fee simple, to use the legal term. Compared to the opportunities offered by New Spain and New France, the Atlantic colonies seemed irresistibly attractive. Little more than a century after the first permanent settlement was established in Virginia, more than one and a quarter million settlers were scattered across the wide, empty spaces between the coast and the mountains.

It was from England that the idea of land as property had originally come, and it was no coincidence that with it had arrrived Gunter's chain—22 precisely calibrated yards—and the habit of showing the estate's exact extent on a surveyor's plat drawn to scale. From England, too, came the philosophical underpinning developed by John Locke that the individual earned a right to property by "mixing his labour" with what had once been held in common, and every landowner who had ever enclosed, manured, and improved a field understood this proposition perfectly. But by the middle of the eighteenth century, the American idea of landed property had evolved beyond its English roots. Americans had begun to speculate on land. It had become a commodity.

In 1751 Benjamin Franklin stated openly what was apparent to authorities on either side of the Atlantic, that the population of the colonies was growing at such speed, it would double to 2.6 million by 1775. It would not be long, he predicted provocatively, before "the greatest number of Englishmen will be on this side of the water." To some Americans, that prospect raised constitutional questions about being controlled from across the Atlantic, but to the planters of Virginia and the Carolinas, and to financiers in New York and Philadelphia, it also indicated that the purchase of American land was a wise investment.

The opportunity was most apparent to surveyors like George Washington and Peter Jefferson. However much they could earn from surveying fees, it was dwarfed by the profits to be made from buying good land cheap. "The greatest Estates we have in this Colony," the young George Washington acknowledged in 1749 after a summer spent surveying the vast Fairfax estates, "were made . . . by taking up & purchasing at very low rates the rich back Lands which were thought nothing of in those days, but are now the most valuable Lands we possess." In 1752, at the age of twenty, he purchased 1,459 acres in Frederick County, in the Virginian piedmont, the first step in a career of land dealing that eventually made him owner of more than 52,000 acres spread across six different states. Washington usually "improved" his holdings by clearing them of trees, but for most speculators their property rights did not depend on any idea of "mixing their labour" with the soil. Their sole claim to ownership lay in the survey and the map that came from it.

It was the attraction of good land going cheap that drew tidewater speculators into the piedmont, but in 1756 a South Carolina surveyor, John William de Brahm, was sent to build a fort at Loudon on the Little Tennessee River on the other side of the Appalachians, in country that still belonged to the Cherokee. "Their vallies are of the richest soil, equal to manure itself, impossible in appearance ever to wear out," he reported back in admiration. "Should this country once come into the hands of the Europeans, they may with propriety call it the American Canaan, for it will fully answer their industry and all methods of European culture, and do as well for European produce. . . . This country seems longing for the hands of industry to receive its hidden treasures, which nature has been collecting and toiling since the beginning ready to deliver them up."

Control of all this desirable territory as far west as the Mississippi River still remained with the French, but in 1763 they were forced to cede it to the British following their defeat in the French and Indian War. Soon other surveyors took the chance to follow de Brahm. Their findings were brought together in a famous map by Thomas Hutchins, not published until 1778, but whose attractions were known a decade earlier.

On the map's crackling parchment, the Appalachians appear as a palisade of sharp pyramids running from the bottom left-hand corner to the top right-hand corner; but west of them are broad rivers and rolling hills de-

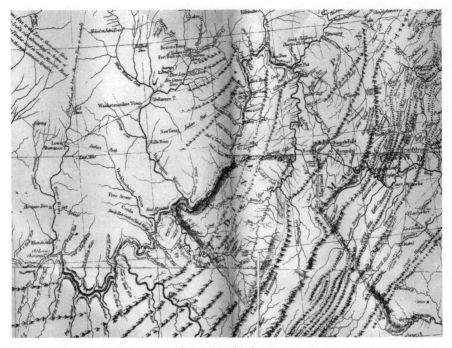

Thomas Hutchins's map.

noted by lines that curl gently toward the Mississippi and are interspersed with Hutchins's own observations in neat italic writing: "A rich and level country"; "Very large natural meadows; innumerable herds of Buffaloe, Elk, Deer, etc feed here"; and along the Wabash River, "here the country is level, rich and well timber'd and abounds in very extensive meadows and savannnahs; and innumerable herds of Buffaloe, Elk, Deer, etc. It yields Rye, Hemp, Pea Vine, Wild Indigo, Red & White clover etc."

That the land belonged to the Cherokee, Shawnee, and Six Nations was a detail that could be overcome by personal negotiation or by killing and terror. To the colonists it was obvious that, with the French claims removed, the entire area between the mountains and the river lay open for occupation. Judge Richard Henderson of North Carolina, who promptly sent surveyors to scout out the land, put into words the dream that drew him and thousands more into the western territory beyond the Blue Ridge and the Al-leghenies. "The country might invite a prince from his palace merely for the pleasure of contemplating its beauty and excellence," he wrote, "but only add the rapturous idea of property and what allurements can the world offer for the loss of so glorious a prospect?"

A profound change had occurred since the first royal charters were is-sued granting the land "in free and common soccage." Many of the original proprietors had gone, and colonial legislatures, such as Virginia's House of Burgesses, were effectively forums for the colonists' interests. It was easy for settlers, squatters, and speculators looking longingly toward the land beyond the Appalachians and enraptured by the idea of property to overlook the king's power.

Then on October 7, 1763, came a harsh reminder of the legal reality be-hind American property. By royal proclamation, George III declared it "to be our Royal Will and Pleasure . . . that no Governor or Commander in Chief in any of our Colonies or Plantations in America do presume for the present, and until our further Pleasure be Known, to grant Warrants of Survey, or pass Patents for any Lands beyond the Heads or Sources of any of the Rivers which fall into the Atlantic Ocean from the West and North West." In effect, a line had been drawn along the watershed of the Appalachians beyond which land could not be measured and owned, and everyone who had already set-tled west of it was commanded "forthwith [to] remove themselves from such settlements."

George III had the right to order it because legally all the land in British America was his. Full in the path of the property seekers was planted the king's feudal authority.

FOUR

Life, Liberty, or What?

THERE WERE MANY STRANDS leading to the moment when the colonists felt driven to weave their anger together into a single declaration of opposition to rule from London. The decision of the British Parliament to close the port of Boston in 1774 as punishment for the destruction of a valuable cargo of tea brought to the surface the resentment of northern merchants already burdened by duties on their goods, a general fury at the earlier killing of civilian rioters by British troops, and a pervasive fear that colonial assemblies were powerless against the king's ministers. But that autumn, when delegates of the discontented colonists convened in Philadelphia as members of the First Continental Congress in order to articulate their grievances, it was not by chance that the first resolution they agreed was "that they are entitled to life, liberty, & property."

In this instance, *property* meant possessions in general, but for Virginians especially it was land that they had in mind, and in particular land beyond the Appalachians. Hence the declaration of the first paragraph of the Virginia constitution, drawn up in June 1776 by George Mason: "That all men are by nature equally free and independent, and have certain inherent rights . . . namely, the enjoyment of life and liberty, with the means of acquiring and possessing property, and pursuing and obtaining happiness and safety." Entitlement to this sort of property was a subject on which the humblest Conestoga mule-driver was at one with the grandest planter.

Barely ten years earlier, the reaction of Colonel George Washington to the royal veto on the acquisition of land beyond the mountains could have served as a warning of what was to come. There had been no more loyal and energetic commander in the French and Indian War, but the colonel was also a Virginia planter and land speculator, and his views were widely shared.

"I can never look upon the Proclamation in any other light (but this I say between ourselves) than as a temporary expedient to quiet the minds of the Indians," Washington wrote in 1767 to his colleague and fellow-surveyor Colonel William Crawford. "It must fall, of course, in a few years, especially when those Indians consent to our occupying those lands. Any person who neglects hunting out good lands, and in some measure marking and distinguishing them for his own in order to keep others from settling them, will never regain it."

Washington was not a man to be deflected even by his sovereign's express command, although as a serving officer he deemed it best to be discreet. "If you will be at the trouble of seeking out the lands," he continued to Crawford, "I will take upon me the part of securing them, as soon as there is a possibility of doing it and will, moreover, be at all the cost and charges of surveying and patenting the same. . . . By this time it will be easy for you to discover that my plan is to secure a good deal of land. You will consequently come in for a handsome quantity."

Alas, Crawford never did. The king's veto was still nominally in force when he was taken prisoner by a Cherokee band while leading a column of troops in territory beyond the Appalachians. On August 3, 1782, the *Virginia Gazette* carried a report of the ordeal of a Dr. Knight, who had been captured along with Washington's colleague:

The unfortunate Colonel was led by a long rope to a stake, to which he was tied, and a quantity of red-hot coals laid around, on which he was obliged to walk bare-footed, the Indians at the same time torturing him with squibs of powder and burning sticks for two hours, when he begged of Simon Gurry (a white

*renegade who was present) to shoot him. [Gurry's] reply was, "Don't you see I
have no gun." [Crawford] was soon after scalped and struck several times on
the bare skull with sticks, till being exhausted, he laid down on the burning em-
bers, when the squaws put shovel-fuls of coals on his body, which made him
move and creep till he expired. The Doctor was obliged to stand by and see this
cruelty performed; they struck him in the face with the Colonel's scalp, saying
"This is your great Captain's scalp, tomorrow we will serve you so."*

Luck was with the terrified doctor, however, and he was able to escape
under cover of darkness.

Gruesome stories such as these were used to justify acts of equal cruelty
on the other side. John Heckewelder, a Moravian missionary living with the
Tuscarawa Indians, told in 1773 of the Indian hunters "who maintained that
to kill an Indian was the same as killing a bear or a buffalo and would fire on
Indians that came across them by the way—nay more, would decoy such as
lived across the river to come over for the purpose of joining them in hilar-
ity; and when these complied they fell on them and murdered them."

These atrocities were evidence of the mounting conflict between land-
hungry colonists and Native Americans. Consequently, when George III
banned land purchases beyond the mountains, it was, as the proclamation
worded it, so that "the several Nations or Tribes of Indians with whom We
are connected, and who live under our Protection, should not be molested or
disturbed." But even Americans who might have sympathized with this strategy could not accept the king's feudal right to impose the ban.

The power that the land beyond the mountains exerted on people's
minds can be deduced from the attempts to circumvent the veto. Two land
companies had been created to speculate in the West before the ban was in
place: the Ohio Company of Associates, which proposed to purchase
500,000 acres beyond the Ohio River; and the Loyal Land Company, organ-
ized by Peter Jefferson with investment coming mostly from his neighbors in
Goochland County, Virginia, which aimed publicly at buying 800,000 acres
of Kentucky but privately had ambitions of exploring and acquiring millions
more as far as the Pacific Ocean. In the fifteen years after George III's proclamation, a whole succession of similar speculative ventures came into being.

The Mississippi Company was created in 1768 to settle land along the river, with Washington as one of its founders; followed by the Illinois and Wabash Company, in which Patrick Henry had an interest; and the Watauga Association, which settled eastern Tennessee and later tried to establish the independent state of Franklin. In 1775 Judge Richard Henderson of North Carolina sent Daniel Boone, a brave scout but an incompetent surveyor, to find territory in southern Kentucky, where he signed a treaty with the Cherokee Indians giving Henderson's Transylvania Company several million acres—Boone's surveying lapses made it unclear exactly how much land was involved. The most ambitious of them all, the Vandalia Company, which aimed to acquire sixty-three million acres in what is now Illinois and Indiana, employed Benjamin Franklin as its London agent and even counted among its members such influential figures in the British government as the future prime minister, Lord North, and the lord chancellor, Lord Camden. "One half of England is now land mad," remarked one of its promoters, "and everybody there has their eyes fixed on this country."

These powerful interests created some loopholes in the prohibition, but it was still in place in 1773 when a young surveyor named Rufus Putnam sailed up the Mississippi River to Natchez. There he began surveying more than one million acres on the banks of the river so that it could be sold to New England veterans of the French and Indian War. Among the mostly southern land speculators, he stood out because he came from Massachusetts, but in any company he would have been noticeable. In character he more or less resembled the coat of arms he later adopted, which showed three bristly wild boars below a roaring lion surrounded by thistles and the motto, in spiky Gothic lettering, "By the name of Putnam." Almost everything he achieved he owed to his ferocious determination.

In 1745, when he was barely seven years old, his father died, leaving the family destitute. His mother's second marriage was to an illiterate drunkard named Sadler. "During the six years I lived with Capt Sadler," Rufus wrote bitterly, "I never Saw the inside of a School house except about three weeks." At the age of fifteen he apprenticed himself to a man who built watermills, and sucked up knowledge where he could, but as he confessed in his autobiography, "having no guid I knew not where to begin nor what corse to pursue—hence neglected Spelling and gramer when young [and] have Suffered much through life on that account."

Putnam's prospects were transformed by the French and Indian War. In 1757 he joined the Royal American Regiment, where he became an engineer, a trade that taught him how to carry out every kind of measurement. When peace came in 1763, the orphan used his new skill first to build mills, and later to survey land, and soon felt financially secure enough to marry the wealthy Persis Rice, daughter of Zebulon. In the first six years of their marriage, she bore him four daughters and a son. What

Rufus Putnam

Putnam wanted, he wanted immediately. Neither in private nor public life did he ever show any guile, and rarely much patience. With five children to feed, and the expectation of more on the way, he turned to land speculation, surveying and acquiring land in the West Indies and what is now Alabama for New England veterans. The Natchez venture was on a far larger scale, and like Washington, who was trying to secure land on the Ohio River for Virginia veterans of the same war, he proposed to keep some of the best sites for himself.

Putnam had spent eight months laying out the ground for future settlements at Natchez when in 1774 the Board of Trade in London declared he was in breach of the royal proclamation and issued an order specifically forbidding any further surveys on the Mississippi. Returning home to Massachusetts in a fury, Putnam found his personal frustration echoed in the accumulating anger of the colonists, and he was one of the first men to be commissioned into the Massachusetts Regiment following the ringing shots at Lexington in April 1775.

Almost at once, he was able to demonstrate his talent for finding practical solutions to complex problems. Following the outbreak of hostilities, an American force of 11,500 men under George Washington surrounded the British army under General Thomas Gage in Boston. With little faith in his untrained men against 6,000 British regulars, Washington needed to pen Gage in, but lacked anyone with the knowledge to construct the necessary siege-works. In the same uncomplicated way that he had taught himself to be a sur-veyor, Putnam read a book on military engineering, then laid out a system of trenches and defensive posts that kept the British cooped up until the spring of 1776, when the threat of artillery bombardment forced them to sail away.

Washington rewarded his bristly subordinate by appointing him chief engineer to the army, and to the end of his days Rufus Putnam remained Washington's man, body and soul. In peacetime, all political questions were solved by doing whatever Washington required, and if that was not clear, Putnam would write to ask. Anyone who opposed his general and later his president was an enemy; and one particularly prominent opponent, whose name Putnam could not bring himself to mention in his memoirs, was an "Arch Enemy."

*E*VEN IN 1776 it was evident that Thomas Jefferson marched to the beat of a different drum. Indeed, if there was any one person immune to the general lust for land beyond the Appalachians, it was Jefferson, the Virginia planter, who, in his wording of the Declaration of Independence, changed the fundamental assertion of rights from "life, liberty and property" to "life, liberty and the pursuit of happiness."

It is one of the greatest paradoxes in Jefferson's paradoxical character that he—who was to acquire more western land on behalf of the United States than any speculator could have dreamed of possessing, who laid the foundations for the nation's further territorial expansion to the Pacific by sending Meriwether Lewis and William Clark to find a route to the coast in 1803, who believed passionately in the virtues of owning land and adored his own plantations and garden at Monticello—was so tepid in acquiring it for himself. From his father, Thomas inherited about 7,000 acres in the Virginia piedmont, including Monticello and its farms, and his marriage to Martha Wayles Skelton in 1772 added to his holding. He even joined several schemes for acquiring land beyond the mountains, but then neglected them, and all eventually failed.

By birth and upbringing, Jefferson belonged to the Virginia plantation aristocracy. His family had been there since the late seventeenth century; generations of Jeffersons had followed the frontier west, and to them land acquisition was second nature. As a child, Jefferson was clearly destined for the aristocratic role. The model before him was that of his large, rawboned father, Peter Jefferson, who had not only run the Fairfax line with Thomas Lewis through the swamps and ravines of the Blue Ridge, but later extended William Byrd's boundary between Virginia and Carolina almost into Kentucky, and only returned home to build a plantation mansion on the frontier and lay plans with his neighbors to acquire still more land through the Loyal Land Company. Before his death in 1757, when Thomas was fourteen, Peter appointed various friends and relatives as guardians to his children, and all but two of these were trustees of the Loyal Company.

What changed the course of the boy's life was the decision of his guardians to send him at the age of sixteen to William and Mary College in Williamsburg. There he fell under the influence of a remarkable teacher named William Small. In his autobiography, written in his seventies, Jefferson confessed, "It was my great good fortune, and what probably fixed the destinies of my life that Dr. Wm. Small of Scotland was then professor of Mathematics, a man profound in most of the useful branches of science, with a happy talent of communication, correct and gentlemanly manners, & an enlarged & liberal mind." Jefferson was famously guarded in his emotions, but where Small was concerned he became almost fulsome. "Dr. Small was . . . to me as a father," Jefferson confided to a friend. "To his enlightened and affectionate guidance of my studies while at college, I am indebted for everything." The brief reference in his autobiography to his real father, the great surveyor, is stark by comparison: "My father's education had been quite neglected; but being of a strong mind, sound judgment and eager after information, he read much and improved himself."

In 1760 William Small was just twenty-four, a product of those Scots universities that bred in the likes of David Hume and Adam Smith a restless desire to find a rational key to understanding human nature and human society, and to sixteen-year-old Thomas Jefferson he must have seemed like an ideal elder brother. "He, most happily for me, became soon attached to me &

made me his daily companion when not engaged in the school," Jefferson remembered gratefully, "and from his conversation I got my first views of the expansion of science & of the system of things in which we are placed."

The most obvious consequence of Small's company was that all his life Jefferson preferred to think of himself as a scientist rather than a politician. "Nature intended me for the tranquil pursuits of science, by rendering them my supreme delight," he wrote soon after his retirement from the presidency, "but the enormities of the times in which I have lived have forced me to take a part in resisting them and to commit myself on the boisterous ocean of political passion." Politics, even sometimes the United States itself, he would refer to as "an experiment." In his house he hung portraits of Francis Bacon, modern science's founding father, and Sir Isaac Newton, its greatest luminary. He studied botany, the most highly developed science of the day, and prided himself on his membership in the American Philosophical Society, the country's leading scientific body. In old age, he boasted to his friend and enemy John Adams, "I have given up newspapers in exchange for Tacitus and Thucydides, for Newton and Euclid, and I find myself much happier."

It was Isaac Newton, above all, who had expanded science and shown how the system of things might be understood. The consequences of Newton's laws of motion were fundamental to the development of physics and astronomy in general. In Newtonian physics there were theoretical explanations for every event, from the movement of the planets to the swing of a pendulum, and two brilliant analyses from his monumental *Philosophiae naturalis principia mathematica* (Mathematical principles of natural philosophy), published in 1687, provided what would prove to be the real starting point for modern measurement. Newton deduced that instead of being a perfect sphere, the earth bulged at the equator and flattened at the poles; consequently, gravity would be stronger at the poles, because they were closer to the center of the earth.

But Newton's laws also taught a lesson that went beyond physics. In the universities of Aberdeen, Edinburgh, and Glasgow, students were encouraged to believe that just as it was possible to discover through rational inquiry the basic principles that governed the natural world, so reason made it possible to understand the principles governing human nature. This, the essential outlook of the movement known as the Enlightenment, was what William Small passed on, and what Jefferson meant by understanding "the system of things in which we are placed."

"Fix reason firmly in her seat, and call to her tribunal every fact, every opinion," he would later advise his young nephew, in an echo of Small's teaching. "Question with boldness even the existence of a god; because, if there be one, he must more approve of the homage of reason, than that of blindfolded fear." Once rational inquiry had uncovered the principles of human nature, it would be possible to establish the conditions under which individuals could be allowed as much liberty as they desired, and yet a rational, self-regulating society would emerge.

In 1764 Small returned to Britain, but Jefferson never ceased to feel for the man who, as he acknowledged, "filled up the measure of his goodness to me, by procuring for me, from his most intimate friend G[eorge] Wythe, a reception as a student of law under his direction, and introduced me to the acquaintance and familiar table of Governor [Francis] Fauquier, the ablest man who had ever filled that office. With him, and at his table, Dr. Small & Mr. Wythe . . . & myself, formed a *partie quarrée*, & to the habitual conversations on these occasions I owed much instruction."

Thomas Jefferson

Recognition of Jefferson's place as an insider came with his election in 1769 as a twenty-six-year-old lawyer to the Virginia legislature's House of Burgesses, which represented plantation owners' interests. Yet Jefferson was never a typical member of his class. He thought about land in a way that no speculator would. In a memorable passage in *Notes on the State of Virginia*, the book written from 1780 to 1782 that expresses some of Jefferson's deepest beliefs, he explained, "Those who labour in the earth are the chosen people of God, if ever he had a chosen people, whose breasts he has made his peculiar deposit for substantial and genuine virtue." Other trades had to depend "on the casualties and caprice of customers. Dependence begets subservience and venality, suffocates the germ of virtue, and prepares fit tools for the designs of ambition." But "corruption of morals in the mass of cultivators is a phaenomenon of which no age nor nation has furnished an example."

It is hard to think of any other Virginian who might have entertained such a far-fetched idea. Not William Byrd, who observed of his fellow planters, "Our land produces all the fine things of Paradise, except innocence." Not Washington, cheated by a farming acquaintance, William Clifton, into paying an extra £100 (about $470) in 1760 for an estate neighboring on Mount Vernon. Not Jefferson's friend Fielding Lewis, who remarked of the land deals in the piedmont that "every man now trys to ruen his neighbour." For them land was a source of wealth, not the basis of God-given virtue.

To Jefferson, however, the possession of land was the Newtonian principle that made a democratic society work. It guaranteed the independence of the individual and gave each one an interest in building a law-abiding community. All that was then needed was education to teach them how best to use their freedom. The ideal he had in mind existed in prefeudal Saxon society, with its local court and administration based on the "hundred," or parish, and its values derived from the stout-hearted, independent-minded yeomen farmers who worked the soil. Consequently, all the political systems he devised, for counties as for nations, shared one fundamental quality: the widest possible distribution of land.

When he began to question the basis of the royal claim to exercise feudal power over the land beyond the mountains, his conclusion was what his class would have expected—that George III had no right to restrict their desire to acquire property. But his argument was more radical than most planters defending their estates against squatters would have found acceptable.

In Saxon England, before William the Conqueror had imposed his regime, Jefferson argued, feudalism was unknown. "Our Saxon ancestors held their lands, as they did their personal property, in absolute dominion," he declared in a fiery pamphlet, *A Summary View of the Rights of British*

America, published in 1774. It was William and his Norman invaders who had invented "the fictitious principle that all lands belong originally to the king," and the fiction had been maintained by his successors. Since America had been occupied and won without help from the Crown, George III had no grounds for claiming power over the disposal of its land. Only a democratically elected legislature had that power, Jefferson concluded, and, in a phrase that would have been music to any squatter's ears, wrote that if it failed to do so, "each individual of the society may appropriate to himself such lands as he finds vacant, and occupancy will give him title."

It was this sweeping attack, cutting at the very foundation of royal power over America, that led to Jefferson's appointment by the Continental Congress in 1776 to the three-man committee responsible for drafting the Declaration of Independence. In his original draft, his onslaught on George III's feudal powers continued for many paragraphs, to the evident confusion of other representatives, who cut them out. But all his life, Jefferson remained in love with the Saxon ideal.

When he returned to the Virginia legislature, after independence was declared in July 1776, he was given an early chance to put his landownership theory into practice. His reputation as the Declaration's prime author led to his appointment to a committee revising the state's laws so that they would reflect republican rather than royalist values. Although the best known of his proposals was the "Statute for Establishing Religious Freedom," a landmark in legislative tolerance, his primary goal was the wider distribution of land. He introduced a bill to give 75 acres to any Virginian who did not already have any land, and to offer a "headright" grant of 50 acres to every landless immigrant who arrived in the state from overseas. Together with Virginia's generous promise of land to soldiers enlisting in its regiments—ranging from 100 acres for enlisted men up to 15,000 acres for a major general—Jefferson's land grant proposals would have created a network of small farms guaranteeing the future health of democracy in the state. In the same spirit, he drew up legislation to change inheritance law so that a landed estate could be left equally to all the children rather than to the eldest son alone, and to abolish entail—or wills preventing heirs from breaking up estates.

To complete the scheme, he came up with a plan for wholesale administrative and educational reform. As he put it, "I drew a bill for our legislature, which proposed to lay off every county into hundreds or townships of 5. or 6. miles square, in the centre of each of which was to be a free English school." Jefferson's high standing in the Virginia assembly led in 1780 to his appointment as governor, but he never succeeded in convincing the legislature of the merits of creating a society of yeoman farmers.

*I*N THE FALL OF 1782, Jefferson's wife, Martha, died at the age of thirty-three after giving birth to their sixth child. They had been married for less than eleven years, and in that time Martha had been almost constantly pregnant, with six live children born, although only two survived beyond infancy, and three other pregnancies that ended in miscarriages. At her death, Jefferson was prostrated by a grief so consuming that for a month he could not face anyone and stayed secluded in a room where, his daughter recalled, he wept and groaned, emerging at last only to go for long, solitary horseback rides in the mountains. There may have been an element of guilt in this—the risk of repeated pregnancies to the health of delicate women was well understood—but whatever the source of his emotions, it was obvious that the force of them was overwhelming. In retreat from the pain of grief, he threw himself into work, which absorbed all his energies and attention.

One year earlier, in October 1781, Lord Cornwallis and his British army, hemmed in by Washington on land and by the French fleet at sea, had surrendered, and American negotiators were in Paris deciding the terms of the peace. At home the Continental Congress, which represented the nearest thing to a central government that the thirteen states could agree on, was attempting to work out the new nation's future government and how to pay off the mountain of debt accumulated in paying for Washington's Continental Army. The United States's single asset, if the negotiators could pry it from Britain's grasp and the states were prepared to give up their own claims to it, was the land between the Appalachians and the Mississippi.

"There are at present many great objects before Congress," wrote the Rhode Island delegate, David Howell, early in 1784, "but none of more importance or which engage my attention more than that of the Western

Country." How the land would be used—sold to pay the country's debts, parceled out to republican farmers, distributed among the existing states, divided up to create new states, handed over to settle the claims of the prerevolutionary land companies—would in large part determine the relationship of the central government to the states.

In June 1783 Jefferson was elected a delegate from Virginia to the Continental Congress and immersed himself in the many great objects before it—above all, in the question of the western lands. Preserved in the Library of Congress are pages of his comments on proposed legislation, and draft bills whose margins are filled with his detailed annotations. In the space of less than a year, from June 1783 to May 1784, this escape into mental work produced numerous contributions to U.S. law, and three measures so substantial that they were to permeate every aspect of American life: the invention of the dollar, the procedure for creating states from the Western Territory, and the means of surveying that territory. Had Jefferson had his way, there would have been a fourth: the invention of a new set of weights and measures. It indicates the cohesion of his thinking that all four formed part of a single logical structure.

FIVE

Simple Arithmetic

FOR A YEAR AFTER WINNING independence at the Treaty of Paris, the United States of America barely existed. By the Articles of Confederation drawn up in 1777, the thirteen original states had entered into "a firm league of friendship with each other," and with a common enemy to face, a national army to support, and a unifying purpose in mind, the new republic required no clearer definition. Once peace came, its identity grew indistinct. Its army, after the demobilization of Washington's Continental Army, consisted of a single regiment. There was a national post office, but there was no national judicial system, and no national executive other than three small departments responsible for defense, finance, and foreign affairs. The Continental Congress had no legislative autonomy except with the agreement of at least nine states, and no powers of taxation. Above all, the United States had no territory.

Throughout most of the war, all the desirable land west of the Appalachians had been claimed by individual states rather than by the union. Under the terms of its 1609 charter, Virginia asserted a right to all the land from Lake Erie in the north to St. Louis in the south, while Massachusetts could point to a phrase in its charter giving it "the mayne Landes from the Atlantick . . . on the East Parte, to the South Sea [the Pacific] on the West parte." In protest Maryland, whose western boundaries had been limited by its charter, refused even to ratify the Articles of Confederation until these gigantic demands were abandoned. In 1781 Jefferson, while governor of Virginia, agreed to give up his state's claim, transferring it to the central government, and when all of the other claimant states except Georgia followed suit, Maryland signed the Articles. By the Treaty of Paris in 1783, the British ceded to the Americans all the western country as far as the Mississippi River, and at that point what had once been George III's land should have belonged to the United States.

True to his Enlightenment self, however, Jefferson had added a reservation. Virginia would cede its claims to the U.S. government alone, he explained, and consequently only that government would have the legal right to acquire ownership of the land from the Native American nations who lived there. This meant that any claims made by prerevolutionary land companies would become invalid. Congress, deeply influenced by company sympathizers, refused to accept the deal, and for a year there was stalemate. Until one side or the other backed down, no U.S. territory could exist.

Among those who grew impatient at the delay was General Rufus Putnam. His rank was a reward for the sterling service he had rendered Washington both at the siege of Boston and during the fighting in New York. Following the Continental Army's strategic retreat in 1776, Putnam had returned home to command the Fifth Massachusetts Regiment in defense of his own state. There he had found time to father three more children, two in the dark days of defeat, and a third to celebrate the approach of victory—not for nothing were there rampant boars on the Putnam coat of arms—and he had acquired the fine estate of Rutland that formerly belonged to a wealthy loyalist. With the return of peace, he resumed the work he had been doing before the war.

In April 1783 Timothy Pickering, a delegate to the Continental Congress, reported that "there is a plan for the forming of a new State Westward of the Ohio. Some of the principal officers of the Army are heartily engaged in it. The propositions respecting it are in the hands of General [Jedediah] Huntington and General Putnam, the total exclusion of slavery from the State to form an essential and irrevocable part of the Constitution."

Putnam and his fellow officers wanted to acquire almost eighteen million acres west of the Ohio River and north to Lake Erie—as much land as any of the prerevolutionary companies—in order to have the entire area settled by veterans of the Continental Army. Most soldiers had received part of their pay in paper money supplemented by military warrants on completion of service, but so many of these notes had been issued they had lost more than three-quarters of their value. Putnam proposed that the warrants be used to pay for the Ohio land at face value. This would have had the double benefit of settling seasoned soldiers in a part of the United States vulnerable to British invasion from Canada, and of reviving the value of the military warrants whose worth he judged to be "no more than 3/6 & 4/- (approximately one dollar) on the pound (just less than five dollars) [but] which in all probability might double if not more, the moment it was known that Goverment would receive them for lands in the Ohio Country." He petitioned Congress to grant him the land, and when Congress did not respond, he wrote Washington again in April 1784 to ask, as was his habit, what should be done. "The Settlement of the Ohio Country, Sir, ingrosses many of my thoughts and much of my time sence I left Camp," he admitted, but the delay was making the veterans impatient. "Many of them are unable to lie long on their oars waiting the desition of Congress on our petition."

Washington, who had returned to his estates after giving up the command of his victorious troops, was no less impatient to have Congress make some decision about the land beyond the mountains. Throughout the war, a stream of settlers had moved westward into Tennessee and Kentucky, and the increase in their numbers after the fighting was over prompted Washington to warn the Congress that without some policy, "the settling, or rather overspreading of the Western Country will take place by a parcel of *Banditti* who will bid defiance to all Authority." The thorny prerevolutionary conflict between settlers and squatters, proprietors and "Goths and Vandals," had not gone away.

Yet nothing could be done until Congress agreed to accept Jefferson's condition for ceding Virginia's claim. There was no chance that Jefferson himself would give way. In 1783, while he was absorbed in the affairs of Congress, Virginia set about disposing of unoccupied land inside its existing boundaries, and his Saxon blueprint was obliterated. The Virginia Assembly was determined to sell the territory with as little restriction as possible, and the legislation authorizing the sale allowed the process to become a swill bucket for speculators. Surveyors were bribed into setting aside the best plots, land warrants were acquired cheaply from army veterans, and wads of the state's devalued paper currency, which carried the right to claim unoccupied land, were bought up for a quarter of their face value. One speculator alone, Robert Morris, acquired one and a half million acres of western Virginia.

It was a lesson Jefferson did not forget. During the next twenty-five years he was to engage in an ideological war with land speculators, whose interests were so diametrically opposed to his, and where the sale of the United States's land was concerned, he demonstrated that he was prepared to do everything in his power to thwart their aims. Time after time, he found himself confronting Robert Morris, the Philadelphia merchant who had profited so richly from Virginia's land sales. In the Continental Congress, Morris held the post of superintendent of finance, an influential position which helped ensure that the congressional mood remained in favor of the land companies.

They were polar opposites: Morris, whose fat, friendly, asthmatic appearance distracted attention from a gambler's cold, abacus mind; and the lean, controlled, complex Jefferson, concealing emotional vulnerability and romantic idealism behind an air of studied informality and a stream of logical argument. "His whole figure has a loose, [shambling] air," observed Senator William Maclay in 1790. "He has a rambling vacant look, and nothing of that firm, collected deportment which I expected. . . . He spoke almost without ceasing. But even his discourse partook of personal demeanour. It was loose and rambling, and yet he scattered information, wherever he went, and some even brilliant sentiments sparkled from him."

Unlike Jefferson's privileged background, Morris's past was one of unremitting struggle, from his arrival as a penniless immigrant from England, through long years as an accountant working for the wealthy Philadelphia merchant Charles Willing, until he was made a partner in Willing's company and became one of the wealthiest men in America. During the war, he had used his wealth to underwrite foreign loans and contracts for the purchase of

Robert Morris

supplies for Washington's army, and with the goodwill these services earned, he secured still more profitable contracts for himself.

The second skirmish in their long campaign occurred over currency. This was one subject for which the Articles of Confederation had given the responsibility to Congress. The basic problem was well illustrated when Washington replied to Putnam in April 1784 on the matter of acquiring land beyond the Ohio. Pointing out that Congress was still deadlocked over the Western Territory, Washington offered instead to lease his own 30,000 acres in the Ohio Valley to the impatient Massachusetts veterans. The annual rental would be high, about thirty-six dollars per 100 acres, he explained, because "it is land of the first quality" and the cost of the improvements he had made amounted to "£1568 Virginia, equal to £1961/3/3d Maryland, Pennsylvania or Jersey currency," and if Putnam was still not sure how much that meant in Massachusetts, Washington added that "a Spanish milled dollar shall pass in payment for six shillings."

The complexity of working out financial values between the different states hobbled every commercial transaction. Although the legal tender remained officially the British pound, divided into twenty shillings, each in turn subdivided into twelve pennies, its value in America differed from one state to the next. The commonest single coin, the Spanish dollar, was worth five shillings in Georgia, but thirty-two shillings and sixpence across the border in South Carolina, and six shillings in New Hampshire, while the official London rate was four shillings and sixpence, added to which it contained eight bits in Pennsylvania but was divided into ten bits in Virginia. Along with the British pounds and pennies, there were French louis d'or and écus, Portuguese pistoles and half-Joes—so called because they carried the image of King Johannes V—and Dutch florins and Swedish dollars, or riksdalers, each with a value that varied from state to state. Familiarity taught most people to juggle the sums, and just as the teenage Washington casually reckoned up his pay in pistoles and doubloons, so Jefferson, scribbling a quick note of a sale of land, recorded that the price had been "200 [pounds] of which 20 half-Joes are paid." The Articles of Confederation had handed over regulation of the currency to the Congress, and this was one area in which the states agreed that the central government should take action.

The first recommendation came from the Congress's superintendent of finance, Robert Morris, and, like his previous actions, mixed public duty with private profit. He suggested a common currency based on a tiny fraction of a penny, and proposed also to be responsible for minting the new coinage, a post that could have been hugely profitable. It was Jefferson who frustrated the project. In a report delivered early in 1784, he argued that Morris had chosen an unrealistically small unit, and recommended instead the adoption of the Spanish dollar as the most convenient basis. In the interests of simplicity, he suggested also that instead of being divided up into eight bits, it should be decimalized. His logic was casual but devastating.

"Every one remembers," he told Congress, "that when learning money arithmetic, he used to be puzzled with adding the pence, taking out the twelves and carrying them on; adding the shillings, taking out the twenties and carrying them on. But when he came to the pounds where he had only tens to carry forward, it was easy and free from error. The bulk of mankind are school boys thro' life. These little perplexities are always great to them." Accordingly, he argued, the dollar should be subdivided into tenths (dismes), hundredths (cents), and thousandths (mills).

It was an argument that everyone could understand, and less than eighteen months later, on July 6, 1785, Congress resolved that "the money unit of the United States of America be one dollar" and that "the several pieces shall increase in decimal ratio." This was not just an intellectual victory. It effectively prevented Morris from achieving his goal of running the U.S. Mint.

But for them both, the land remained the critical issue. Before the war, Morris had set up the North American Land Company, with the legal help of Patrick Henry. Through its interests in other prerevolutionary land companies, it could claim several million acres in the western country, and so Morris had continued to hold up congressional acceptance of Virginia's cession with its awkward conditions. Jefferson, on the other hand, was already working out how the United States should sell its land so that the speculators would not benefit, and how the settlers there should become U.S. citizens. The answer to both questions involved decimals.

Before the land could be sold, it had to be measured, and on this apparently straightforward issue, Morris and Jefferson were again on opposite sides. In the course of the currency debate, Morris had declared "it is happy for us to have throughout the Union the same Ideas of a Mile and an Inch, a Hogshead and a Quart, a Pound and an Ounce." This was stretching the truth, for American measures were almost as confused as the currency. Nevertheless, Morris was against change. Even without his distaste for the financier, it was clear to Jefferson that the rationale for replacing pennies and shillings with a decimal unit applied equally to American weights and measures.

Officially, each state had adopted the system of lengths and troy and avoirdupois weights that Elizabeth had imposed on sixteenth-century England and that subsequent legislation in London had amended. In reality, barely a single unit was the same from one state to the next—except for Gunter's chain. A Virginia tobacco grower like Thomas Jefferson measured his crop in hogsheads, well aware that a Virginia hogshead was larger than a New York hogshead but smaller than one from Maryland, and that a tobacco hogshead from any state was a different size from a brewer's hogshead. A Boston brewer might also refer to his hogshead of beer as a pipe, butt, or puncheon, knowing that each of them contained 2 cooms, 4 kilderkins, 8 rundlets, or 64 gallons. But a Baltimore brewer who used the same measures somehow ended up with only 63 gallons of beer in his Maryland hogshead, while the number of gallons in a Pennsylvania brewer's hogshead actually changed depending on where the beer was sold, because the law required innkeepers to sell beer inside the inn by the wine gallon, which was smaller than the beer gallon that had to be used for selling beer outside the inn. And the confusion over liquid measurements was nothing compared to the labyrinth of quarts, gallons, and bushels used for measuring corn or flour. Because of flaws in English legislation, each of them could be one of eight different sizes and might either be measured heaped above the brim of the container or struck, meaning level with the brim, as custom or the local market dictated.

The direction of Jefferson's ideas can be found on a sheet of paper dating from the winter of 1783–84 and headed innocuously "Some Thoughts on a Coinage," which shows that he conceived of the dollar and a new, decimal American set of weights and measures as two parts of a single system. The weights would be coordinated with the dollar so that a pound would weigh precisely 10 dollars. The lengths were to be derived from the size of the earth.

The idea that the earth might serve as a scientific basis for a system of measures had first been put forward by the French astronomer and cartographer Jean Picard in 1671. The circle of the equator can be divided into 360 degrees, and each degree is subdivided into sixty minutes. The distance of one of those minutes was equal to one nautical mile, a unit that navigators had used since the sixteenth century, and which remains in use today by pilots, mariners, and other navigators. Picard's successor, Jacques Cassini, had estimated the total distance around the equator to be more than 25,000 miles,* which made each degree a little less than 70 miles.

Notes and tables ran down the page, followed on a separate line by Jefferson's calculation for the length of a minute, or "geographical mile"; in his words, "Then a geographical mile will be of 6086.4 feet." Realizing the difficulty of physically measuring the equator, he went on to suggest a way of checking the length of this new mile. "A pendulum vibrating seconds is by Sr I Newton 39.2 inches."

It was Galileo, allegedly dreaming in church and watching a chandelier as it swung slowly from the ceiling at the end of a long chain, who first realized that the length of a pendulum must determine the amount of time it took to move through its arc from one end to the other. The longer the pendulum, the more time it needed to complete the arc. The math describing the phenomenon was provided by Sir Isaac Newton. His calculations showed that in London a pendulum 39.1682 inches long would take exactly one second to swing through its arc. (Because gravity accelerated the swing, the pendulum's length increased fractionally nearer the pole where the pull of

---

*Today's best figure puts it at 24,902 miles or 40,075 kilometers.

gravity was stronger, and shortened nearer the equator.) Jefferson proposed using this scientifically testable unit—known variously as a second, or second's, pendulum—to check the length of his mile. Just as the dollar had been subdivided into dismes, cents, and milles, so Jefferson's geographical mile would be divided into furlongs, chains, and paces equivalent to tenths, hundredths, and thousandths of a mile.

By the time he started to compare his new decimal lengths with traditional units, the geographical mile had already become the American mile in his mind:

Then the American mile = 6086.4 f[eet].   English = 5280 f[eet].

furlong = 608.64 f[eet].   = 660 f[eet]

chain = 60.864 f[eet].   = 66 f[eet]

pace = 6.0864 f[eet] fathom   = 6 f[eet].

The widest discrepancy was with the English mile, but no doubt to comfort himself he listed all the other miles in use, from the Russian—barely 1,500 old yards—through the Irish, Polish, and Swedish, to the Hungarian, which stretched for almost 7 old miles. In such company there was nothing strange about the American mile. There the "Thoughts" ended, a remarkable race through what was evidently a vast fund of knowledge stored in his mind.

It was this new system that Jefferson intended to apply to the most important subject facing the United States: the disposal of the Western Territory.

*I*N THE SPRING OF 1784, resistance to Virginia's condition finally crumbled in the Continental Congress. The national government had debts estimated at forty million dollars, and an income made up of foreign loans and contributions from the states amounting to $4.5 million. Congress was unable to impose taxes, and the states refused to permit it to raise money by putting tariffs on foreign goods coming into the country. The only possible other source of finance was the sale of land. Faced by this grim reality, Morris's supporters deserted him, and Congress voted to accept Virginia's offer. On March 1, 1784, Jefferson led his state's delegation in formally handing over its claims to the immense region west of the Ohio River. For the first time the United States had a territorial reality to match the spiritual identity outlined in the Declaration of Independence and the martial character shown in the Revolutionary War.

On the same day and as part of the deal, a committee chaired by Jefferson produced a report on how this Western Territory was to be governed. It covered each stage of the process, starting with the land's acquisition from the Indians by the U.S. government, through its division into different territories, the choice of territorial government, and their eventual admission as states to the United States. Once the land had been acquired and surveyed, settlement could take place, and when the population of a territory reached 20,000, it could begin the process of becoming one of the states of the Union on a level of equality with the original thirteen founders. There would be no slaves and no hereditary titles in this vision of the expanding republic.

Even the names of some of the proposed states were specified, among them Michigania and Illinoia, which more or less survived, and Assenisipia and Polypotamia, which did not. What was striking was their shape. Except for river and lake boundaries, all were defined by parallels of latitude running east-west—most were to be two degrees deep—and by meridians of longitude running north-south. The Atlantic seaboard states, which had given up claims to the territory, also had their western borders chopped off straight on a meridian. Consequently, the future shape of the United States would not be long and thin, but for the most part square and geometric.

The next day Jefferson was appointed chairman of a committee to choose the best way of surveying and selling off the land that would make up those states. To prevent speculators from acquiring the best lands by bribing surveyors and land registry officials, the committee ruled out the Virginia method of metes-and-bounds surveys with its irregular shapes and complicated registration procedures. Instead, the country was to be surveyed before occupation and divided up into simple squares aligned with each other following the New England model, so that no land would be left vacant. At Jefferson's insistence, these squares were to be called "hundreds," and their sides were to run due east and west, and north and south. Like the shape of the new states, western property was to be rectangular.

To those who have not read "Some Thoughts on a Coinage," the sentence of the committee's ordinance that detailed the dimensions of those squares must appear inexplicable: "[The Western Territory] shall be divided into Hundreds of ten geographical miles square, each mile containing 6086 feet and four tenths of a foot, by lines to be run and marked due north and south, and others crossing these at right angles. . . . these hundreds shall be subdivided into lots of one mile square each, or 850 acres and four tenths of an acre, by marked lines." Since there was no attempt to explain why these particular lengths had been chosen, Jefferson's fellow legislators may not have realized that as well as the opportunity to raise money by the sale of land, he was offering the United States the chance to have the first decimalized system of measurements in the world.

In a letter written a few days later to an old friend, Francis Hopkinson, Jefferson gloated over the audacity of it. "In the scheme for disposing of the soil an happy opportunity occurs for introducing into general use the geometrical mile in such a manner as that it cannot possibly fail of forcing it's way on the people," he began. There would be objections, he acknowledged. Legislators would argue that the report could not be passed into law because it bore "some relation to astronomy and to science in general, which certainly have nothing to do with legislation." He imagined crusty conservatives advocating the preservation of both the 12-penny pound and the 12-inch foot in order to "preserve an athletic strength of calculation." But, he predicted, all opposition was bound to fail: "This is surely an age of innovation, and America the focus of it!"

In those first years when the republic was not yet formed, one quality that marked its leaders was the certainty that all their actions helped give it shape. John Adams thought of it in terms of making a watch. "When I consider . . . that I may have been instrumental in stretching some Springs and turning some Wheels," he wrote to his wife, "I feel an Awe upon my Mind which is not easily described."

Jefferson saw himself as the architect of a self-regulating, land-owning democracy, and in *Notes on the State of Virginia* he outlined his goal. "The proportion which the aggregate of the other classes of citizens bears in any state to that of its husbandmen [farmers]," he stated, "is the proportion of its unsound to its healthy parts, and is a good-enough barometer whereby to measure its degree of corruption. While we have land to labour then, let us never wish to see our citizens occupied at a work-bench, or twirling a distaff." Democracy depended on getting the land into the hands of the people—"It is the manners and spirit of a people which preserve a republic in vigour," he wrote—but unless the transfer was kept simple, Jefferson felt, the speculators would subvert the process to their own ends. Because no shape could be simpler than the square, and no calculation more straightforward than in 10s, the land had to be surveyed in rectangles and measured in decimals.

That, at least, was the rational, scientific conclusion. Rufus Putnam, who had uncoiled his surveyor's chain over miles of New England and West Florida, as well as on the banks of the Mississippi, voiced a different argument. He hated speculators no less than Jefferson. "I am much opposed to the monopoly of lands and wish to guard against large patents being granted to individuals," he told Washington when describing his plan to put veterans on the west bank of the Ohio; "it throws too much power in the hands of a few." He too wanted squares, but his solution was to divide the eighteen million acres into "756 townships of six miles square" and settle communities in each township on the traditional New England model.

The advantage of a square measuring 6 traditional miles by 6 was its ease of subdivision for someone using Gunter's chain. Each side measured 480 chains, a number that could easily be halved, quartered, and so on, according to demand. In fact, the most convenient subdivision split the large square into thirty-six smaller squares, each measuring one square mile. Since one square mile contained 640 acres or 6,400 square chains, a quantity that could in turn be halved no fewer than seven times and still produce a whole number—5 acres or 50 square chains—the practical advantages for surveyors, especially for those with a shaky grasp of Euclid, was obvious.

Only Jefferson could act as an advocate for his plan, because he alone could explain how decimals and democracy and landownership all went together. But in May 1784, before he could present the committee's report to Congress in person, he was appointed U.S. envoy to France. In his absence, William Grayson of Virginia was appointed chairman of the public lands committee. Changes were made, and the decimal measures were the first to go. What remained from Jefferson's original ordinance was the grid pattern of squares with the east-west lines cutting the north-south at right angles. The dimensions that emerged, however, owed more to Edmund Gunter than to Thomas Jefferson.

In the ordinance that Congress passed on May 20, 1785, for "disposing of lands in the western territory," it was laid down that "the surveyors shall proceed to divide the said territory into townships of 6 miles square, by lines running due north and south, and others crossing these at right angles, as near as may be. . . . The lines shall be measured with a chain; shall be plainly marked by chaps [blazes] on the trees, and exactly described on a plat."

The 36-square-mile townships were to be divided into 1-square-mile lots, four of which in each township were reserved to the government "for the maintenance of public schools." Every alternate township was to be sold whole and the intervening ones by square-mile lots—now for the first time called "sections"—at a price of one dollar an acre. The surveyors were also to make note of prominent features like salt licks, mines, mills, mountains, and the quality of soil, and their compasses were to be adjusted to due north.

Putnam was delighted by the proposals, which were almost identical to those he advocated, but in a report to the absent Jefferson, James Monroe commented discreetly, "It deviates I believe essentially from your [recommendations]." Nevertheless, the measuring of America could at last begin.

SIX

A Line Drawn in the Wilderness

THE POINT OF BEGINNING had been decided by a boundary commission headed by two of the finest surveyors in the United States: Andrew Ellicott, who would later help to lay out the plan for the nation's new capital on the Potomac; and David Rittenhouse, whom Jefferson declared to be the greatest astronomer in the world. Their task was to mark out the western boundary of Pennsylvania, running it north across the Ohio River and then on toward Lake Erie. That boundary had been specified in the original charter to William Penn. Until Virginia ceded its claims to the Western Territory, everything beyond that limit might theoretically be part of Virginia—all of Kentucky, most of present-day Ohio, Indiana, and as far west as Wisconsin. Once the line had been drawn, it would become part of the Northwestern Territory.

The caliber of the boundary commissioners was shown by their use of a zenith sector that Rittenhouse and Ellicott had designed and constructed, as well as a theodolite and quadrants. There was never going to be any doubt about the accuracy of the western boundary of Pennsylvania. On August 20 the party reached the Ohio River, and one of the surveyors, Andrew Porter, recorded: "This morning continued the Vista [the line cut through the trees] over the hill on the south side of the River and set a stake on it by the signals, about two miles in front of the Instrument [the theodolite], brought the Instrument forward and fixed it on a high post, opened the Vista down to the River, and set a stake on the flat, on the North side of the River." It was this stake that marked the starting point of the land survey.

From a surveyor's point of view, open forest was preferable to scrub, where every yard had to be hacked clear so that the chainmen could measure out the land. Nevertheless, of the forty-strong team that camped with Thomas Hutchins beside the Ohio River, the largest group consisted of heavy-shouldered axmen recruited to cut down trees that might obscure the view. Since the land belonged to the United States, each of the thirteen states was supposed to have sent a surveyor to take part in the measurement, but only eight had bothered to make the two-week journey by coach from Philadelphia to Pittsburgh over tracks pitted by potholes and tree stumps, and then by horseback to the river. Significantly, at least three of them were advance men for land speculators.

Hutchins must have seemed the ideal choice as leader. He was a good enough surveyor to be part of the Pennsylvania boundary team, and his maps were prized by soldiers and by other cartographers. In his military career, senior officers had repeatedly recorded his sense of initiative, and as a mapmaker he had traveled west of the Mississippi and as far south as New Orleans, seeing more of the continent than almost any other English speaker alive. Everywhere he had shown an abundance of that indispensable frontier quality, self-reliance. As it turned out, however, he was the wrong man for the job.

Temperamentally, Hutchins was a loner, drawn to the unexplored wasteland by its emptiness. Even his obituarist could not help remarking on his "unconquerable diffidence and modesty," which was so marked that when Hutchins was orphaned as a teenager he could not bring himself to ask relatives for help. Instead he ran away to join the army, where anonymity came easily. Showing an aptitude for numbers, he secured promotion first as paymaster then as an engineer preparing plans and fortifications for his unit, the Royal American Regiment, during the French and Indian War. Unlike more sociable colleagues, he devoted his time to maps and notes, and in that thoughtful vein contributed an appendix in 1765 to a regimental war memoir that has become more famous than the main work.

In it he outlined a model for the construction of frontier settlements, recommending that they should be made up of 1-mile-square parcels of land arranged around a square defensive hub on a riverbank. Some historians have given Hutchins credit for thus creating the blueprint for the land survey, but his claim is no stronger than Putnam's, or of those who advocated keeping things simple for dumb surveyors. Squares solved many contemporary problems, but long after the last settler was scalped and the last surveyor learned trigonometry, it was the ideas embodied in the squares that continued to make them powerful. And the most potent was Jefferson's idea that the squares would provide a framework around which democracy could grow.

Writing up his observations of Louisiana, where he was stationed after the French and Indian War, Hutchins described a land of plenty, where German plantation owners grew "indigo, cotton, rice, beans, myrtle-wax and lumber," and the French "employed themselves in making pitch, tar, and turpentine, and raising stock, for which the country is very favourable." As in the North, it was the woodland that was the country's chief glory. "The quantity of lumber sent from the Mississippi to the West India islands is prodigious," Hutchins observed, "and it generally goes to a good market." He presented a vision of riches not simply for information but deliberately to encourage his fellow countrymen to move into this rich territory. "If we want it," he concluded stirringly, "I warrant it will soon be ours."

When the American Revolution broke out, Hutchins was in London overseeing the printing of maps based on his travels. He still held a commission in the British army, and senior officers not only expected him to return to active service but offered him a promotion. It was the defining moment of his life. Hutchins turned down the offer and adamantly refused to fight against his own people. Even when he was arrested, he preferred to let his maps, money, and commission be confiscated rather than change his loyalties. At length he was released and in 1780 found his way back to the United States with a letter from Benjamin Franklin testifying to his unshakable nature. That year George Washington appointed him geographer to his southern army, and in 1785, having demonstrated his surveying abilities as a boundary commissioner for Pennsylvania, Hutchins was appointed geographer to the United States.

In more than forty years of adult life, integrity had been the hallmark of everything that Thomas Hutchins did. But over the next two years, as leader of the surveying expedition, he displayed a degree of ineptitude that could only have come from the incompatibility of his temperament with the task he had been given.

*U*SING A SEXTANT TO check his position against the sun and the polestar, Hutchins established that the survey's starting point lay on the latitude of forty degrees, thirty-eight minutes, and two seconds north.* From that spot the surveyors were to proceed due west, climbing up from the river, working through the choppy hills and valleys of the Allegheny plateau, until they reached the open, level meadows Hutchins's own maps had described as lying beyond the mountains. With a compass, Hutchins took bearings on a distant mark, then sent the axmen forward to clear a path for the chainmen.

The task of the chainmen was described in Robert Gibson's 1739 *Treatise of Practical Surveying*. The rear man stood by the starting stake with one end of the chain, while the front man, carrying the other end and a set of tally pegs, walked toward the mark, unrolling the chain as he went. The *Treatise* stressed the importance of the rear man ensuring that the front one was always in line with the mark. "If the hinder chainman causes the foreman to cover the object," Gibson wrote, "it is plain the foremost is then in a right line towards it. The inaccuracies of most surveys arise from bad chaining, that is from straying out of the right line."

At the end of 22 yards, a tally peg was inserted, the rear chainman came up, and the process was repeated. Ten chains made a furlong; 80 chains made a mile; 480 chains made one side of a township. At each mile, they put in a marker post. The axmen who accompanied the chainmen chopped away trees and bushes, leaving a long trail behind them. As they moved forward, the sur-

---

*Modern methods indicate that he was actually twenty-five seconds, or about 850 yards, farther north than he calculated—an acceptable error for a surveyor of his day who was not using a zenith sector.

veyor checked the position by taking compass bearings on a tall tree or an exposed hilltop, then blazed the tree or marked the hill, and entered the details in his notebook. Thus they progressed through the virgin forest like a caterpillar, hunching up and stretching out, drawing a straight line to the west.

In fact, the land was far from virgin. It had been worked and occupied for centuries by people like the Delaware, Ottawa, Shawnee, and Miami Indians, who alternated between summers farming in the forest and winters buffalo hunting in open country farther west. Pressure from white settlers attempting to cross the Ohio, and from the Iroquois who claimed much of the territory, had pushed a dozen tribes from as far away as Wisconsin to form themselves into a loose alliance known as the Western Confederacy.

The contrast between the surveyors' straight line and the Indians' winding trails and inconspicuous farm plots could not have been more stark. "We do not understand measuring out the lands," a Shawnee spokesman later declared at an abortive peace conference. "It is all ours. Brothers, you seem to grow proud because you have overthrown the king of England." Experience had taught the Delaware and the Miami who lived closest to the Ohio River that British and American attitudes toward ownership were radically different from their own. They sensed the danger in the surveyors' straight line. Permission for the survey had not come from them but from the Iroquois, who, in 1784, signed a treaty at Fort Stanwix, ceding to the United States the territory they claimed west of the Ohio. As the survey inched forward, the Delaware and the Miami watched with growing anger.

Hutchins's party was aware of the hostility and proceeded cautiously. Barely a week after starting, a messenger brought word that an American trading post in a nearby Delaware village had been raided. The news, as Hutchins heard it, was that the trader had been killed, and "all the signs of war were left behind, [the Indians] having marked the inside of the door and contiguous trees with red Paint."

For his nervous team, it was warning enough. On October 8, 1785, they retreated all the way back to Pittsburgh. They had surveyed no more than 4 miles. Hutchins's report made much of the dangers and difficulties they had encountered, particularly the problem of establishing true, as opposed to magnetic, north due to local variations in the magnetic field. Nevertheless, it was an inglorious beginning, and the surveyors, who were being paid at a rate of two dollars a mile, could not have been happy. Congress sent Hutchins back the next year with instructions to run his line westward for only seven squares, or ranges—a distance of just forty-two miles—and, disregarding the land to the north of it, to concentrate on surveying southward to a bend in the Ohio River.

In 1786 the party was strengthened by more surveyors, including notable figures like Israel Ludlow and Winthrop Sargent, both of whom were acting for the Ohio Company of Associates, a land company that intended to buy in the area. With a military guard provided by General Josiah Harmar's troops to keep the Indians at bay, the work went on more quickly. Every 6 miles, a new survey party started running a line due south, and a fresh east-west line was begun 6 miles below the Geographer's Line, as Hutchins's baseline was known. Yet by the winter only half the area that Congress had expected to have available for sale had been surveyed.

The Seven Ranges survey was not completed until June 1787, and the work was not only late but shoddy. The Geographer's Line did not run due west but drooped about two degrees to the south. Few of the east-west lines crossed the north-south ones at right angles, and as a result the intended grid of squares had become a grid of diamonds and irregular quadrilaterals. "The surveyors were apparently all individuals," C. Albert White commented dryly in his magisterial work *A History of the Rectangular Survey System*, "with individual concepts of how to comply with the 'six miles square' and 'right-angle' requirements."

There had to be a reason for work so lamentably below his usual standards, and one was clearly that Hutchins could not establish authority over his crew. But there may have been another. The government's team faced competition from two private land companies, the Ohio Company and an individual speculator, John Cleves Symmes. At least three of Hutchins's surveyors, Ludlow, Sargent, and Absalom Martin, were agents for his competitors, and it was not in their interests that the government should be selling land before their clients were ready.

The speculators had not gone away following the cancellation of their prewar claims. The old assumption that a rising population would push up prices still held good, and for the financially sophisticated a more immediate source of profit existed in the form of the military warrants and other forms of paper currency. These had been issued by both the Continental Congress and individual states during the war, and financial constraints in 1786 forced many states to revert to the practice. As Putnam had noted, so many had been printed that their real worth was reduced to barely a quarter of their face value, but now that Congress proposed to sell land, it was clearly under some kind of obligation to accept in payment its own currency at full value. This arrangement enabled a speculator to pay for 640 acres priced at $640 with warrants bought for $160 or less, guaranteeing an instant profit when the parcel was sold. Robert Morris had used the technique when he made a killing on the Virginia land sale, and he and others now wanted to extend it to U.S. land.

The only question was whether the Congress could be persuaded to ignore the clear intention of the 1785 ordinance to sell land in the Northwestern Territory to individuals and communities but not to speculators. That doubt was dispelled by the Ohio Company, whose originator and guiding spirit was the redoubtable Rufus Putnam. The company was to make him rich, but that was not his original intention. It started as the realization of the idea he had outlined to Washington, to settle New England veterans in the strategically important Western Territory, but soon after the company was set up in January 1786, the nature of this uncommercial plan changed. Among its 800 shareholders were indeed many soldiers from New England, but they also included speculators like Morris, and much of the company's capital of $250,000 was used to purchase military land warrants and depreciated currency on the open market for as little as ten cents on the dollar.

This financial subtlety does not sound like Putnam's style, for everything about him was utterly straightforward. The credit, therefore, must go to his colleague and cofounder, the serpentine Reverend Manasseh Cutler, pastor of the Congregational church in Ipswich, Massachusetts—"stately and elegant in form, courtly in manners," as an acquaintance described him, "and at the same time easy, affable, and communicative. He was given to relating anecdotes and making himself agreeable." It was Cutler, exquisite in black velvet suit, white linen collar, and silver-buckled shoes, who con-

vinced Congress to disregard the pro-
visions of the 1785 ordinance relating
to the survey and division into
squares before sale. It was Cutler who
persuaded Congress to sell the Ohio
Company more than one million
acres immediately, just downriver
from the Seven Ranges, with an option on almost four million more.
Most significantly for the profitability of the scheme, it was Cutler who
made Congress agree to accept payment in devalued military warrants.
His task was made easier by the slowness of Hutchins's survey of the Seven

Manasseh Cutler

Ranges and Congress's impatience to make money quickly from land sales.

The government's price for the land was nominally one dollar an acre,
but Cutler's persuasive manner eventually was to bring it down to twenty-
five cents an acre, most of which was paid in paper money, so that the real
cost to the company was approximately twelve cents an acre. Ostensibly the
company agreed to buy 1.5 million acres to be paid for in four installments,
but a secret deal gave Cutler an option on an additional 3.5 million acres on
which he hoped to make further gains. On the day that the deal was struck,
in July 1787, the entry in the reverend's journal was a decidedly unspiritual
salute to "the greatest private contract ever made in America."

The Ohio Company was a hybrid whose goal was partly to speculate and
partly to settle the land. If Cutler was responsible for the commercial lobbying, Putnam must take the credit for the more idealistic parts of the legislative framework within which the settlement was to live and work.

On July 13, 1787, "an Ordinance for the government of the Territory of
the United States northwest of the River Ohio" was passed by Congress. Between three and five states were to be carved from the Northwest Territory,
and they eventually included all of present-day Ohio, Indiana, and Illinois
and most of Michigan. Their citizens would be guaranteed elected govern-

ment as in the existing thirteen states, but, three years in advance of the U.S. Constitution, they also were assured of freedom of religion, trial by jury, and due process of law. At the last moment a clause was added requiring section 16 in each township to be set aside for the support of public education, on the grounds that "religion, morality and knowledge being necessary to good government and the happiness of mankind, schools and the means of education shall forever be encouraged." As a result of Putnam's own firm convictions, slavery was also banned, almost eighty years ahead of its prohibition by the Thirteenth Amendment to the Constitution.

Much of the rest of this remarkable document bore the imprint of Jefferson's 1784 report and his vision of the way American democracy should develop. Consequently, it began with a section that effectively made feudal land tenure and laws of primogeniture and entail illegal in the territory northwest of the Ohio. Emphasizing that the new land belonged to the United States, it stated specifically that the territorial legislature "shall never interfere with the primary disposal of the soil by the United States in Congress assembled, nor with any regulations Congress may find necessary for securing the title in such soil to the *bona fide* purchasers." Finally, all those exercising authority in the territory, its governor, judges, elected representatives, and voters, were required to be landowners themselves.

For the first time, an American state was to grow up around an entirely American structure of government and an American set of principles. Fixer and dealer that he was, Manasseh Cutler recognized this as the crucial attraction that the Northwestern Territory offered and, in the Ohio Company's prospectus, presented it in characteristically seductive fashion: "There will be one advantage which no other part of the earth can boast and which probably will never again occur—that in order to begin *right*, there will be no *wrong* habits to combat, and no inveterate systems to overturn—there is no rubbish to remove before you can lay the foundation."

It is almost impossible to overstate the significance of the 1787 ordinance, whose emphasis on democratic government and legal institutions was to exert a deep influence on the development of the whole of the Midwest, even beyond the limits of the Northwestern Territory. Yet the paradox was that it sprang out of the commercial needs of a land company. Although Jefferson never wavered in his belief that the speculators' intention was simply "to dab-ble in federal filth"—depreciated currency—and thus make money at the ex-pense of those who actually wanted to work the land, the ordinance suggested a contrary thought. It was possible that speculation and democracy might not be the natural enemies that Jefferson perceived them to be.

In September 1787 the Seven Ranges, the lands surveyed by Hutchins, were at last put on sale in New York City. Fewer than 100,000 acres were actu-ally paid for, and only $117,108 reached the U.S. Treasury. There were good reasons for this disappointing result. The smallest parcel of land on offer, a 1-square-mile section, cost $640, a sum that had to be paid on purchase and was beyond the means of all but wealthy individuals. Although Hutchins's report declared the soil to be excellent—"The Land is too rich to produce Wheat," he wrote, "but it is well adapted for Indian Corn, Tobacco, Hemp, Flax, Oats, etc."—there were grave doubts about the quality of the land on sale. George Washington, who knew the Ohio Valley well and owned 30,000 acres on the east side of the river, thought the ground was too hilly. There was workable soil at the bottom of the valleys, but the rest he judged to be useless except "to support the Bottoms with Timber and Wood." Without personal knowledge of the country, picking the right square in the grid be-came a lottery. Some squares might have no water supply, while others could consist of nothing but swamp. In pessimistic tones Washington warned that "the lands are of so versatile a nature, that to the end of time they will not by those who are acquainted therewith be purchased, either in Townships or in square miles."

Disappointed by the returns, the Continental Congress suspended the public survey. For the next eight years, until 1795, private land companies were virtually alone in selling U.S. land west of the Ohio. Jefferson's great ex-periment appeared to have failed. Even at the public land sale in New York, it was noticeable that, despite all precautions, more than half the purchases were made by a pair of speculators, Alexander Macomb and William Edgar, associates of Jefferson's bête noire, Robert Morris.

Nevertheless, the unique advantage of the grid system was demonstrated when the very first patent was issued at the New York City land office on March 4, 1788. It went to one John Martin, who paid $640 for a 1-square-mile section of the Northwestern Territory, identified as Lot 20, Township 7, Range 4. Even now, this is an almost anonymous parcel of field, wood, and rock in Belmont County, Ohio, but then it was simply part of the forest. Once it had been surveyed and entered on the grid, however, it could be picked out from every other square mile of territory, and be bought from an office 300 miles away on the coast. Whether that gain was enough to outweigh the grid's disadvantages was a matter still to be decided.

In the spring of 1789, soon after the failure of the survey for which he had been responsible, Hutchins died. Neither his character nor his career had equipped him to deal with devious business practices—as one of his obituaries put it, "all join in declaring him to have been 'an Israelite indeed, in whom there is no guile'"—and it is possible that he never realized that his work had been sabotaged. Disheartened, he had left the Seven Ranges at the end of 1786 to work on running the boundary between Massachusetts and New York, and returned only once to the Ohio Valley. There, ironically, he was employed by the Ohio Company to map the banks of the river, and when the company's sales pamphlet appeared with lavish descriptions of the agricultural, mineral and timber riches in the area, it boasted a valuable endorsement from the geographer of the United States: "*I do Certify* that the facts therein related," Hutchins wrote, "are judicious, just and true, and correspond with observations made by me during my residence of upward of ten years in that country."

An affectionate obituary in the *Gazette of the United States* paid tribute to Hutchins's upright character, ending with the poignant phrase "He has measured much earth, but a small space now contains him." With him appeared to have been buried the public survey. All its activities had been taken over by private enterprise.

*I*N THE SPRING OF 1788, Rufus Putnam led the first party of fortyseven settlers to the company's new land. Having built a great barge, which they named the *Adventure Galley*, and several smaller boats, they launched them on the Youghiogheny River in western Pennsylvania and drifted on the current until they met the Ohio at Pittsburgh. From there, they floated silently downstream, past the clearing in the forest made by Thomas Hutchins's surveyors, until on a dank, foggy morning a lookout suddenly saw on the right bank the dim outline of a wooden stockade. This was Fort Harmar at the mouth of the Muskingum River, where the company's purchase began. Frantically, they rowed and poled for shore, but the river carried them past, and the party's first landing was some distance below the intended spot, forcing them to haul the *Adventure Galley* back upstream from the bank. The plaque that today commemorates the place where the settlers first set foot on shore quite properly overlooks this mishap and places the start of the great experiment at the city of Marietta itself, where the Muskingum joins the Ohio.

That it was an auspicious spot was suggested by the presence of tall mounds, part of the Hopewell civilization, constructed there centuries earlier by Native Americans. Between the mounds and the Muskingum, and about three-quarters of a mile away from the Ohio, Putnam ordered the construction of a wooden fortress, surrounded by a palisade, which he named after the god of war, Campus Martius. The settlers were right to be wary. A party of Delaware, camped nearby in the long clover and buffalo grass while they traded buffalo and beaver pelts with the soldiers in Fort Harmar, served as a reminder that not everyone believed this to be Ohio Company land.

Influence within the company was shared by three individuals, and the shifting balance of power can be deduced from the changing name of the town that was planned. It began as Castrapolis, "the armed city," which was Putnam's choice, and for obvious reasons was quickly replaced by Cutler's more agreeable selection, Adelphia, or "brotherhood." This was discarded when the company's secretary and former surveyor, Winthrop Sargent, suggested that French immigrants might become purchasers, and this commercial consideration led to its final name, Marietta, a tribute to the French queen Marie Antoinette. The city was quickly laid out on the banks of the Muskingum.

The original design, drawn in the calm of Massachusetts, was for sixty large squares with sides 120 yards long, each square containing twelve house lots; but these were awkward dimensions for Gunter's chain to measure out on the ground. In the town plan of 1837 the squares have become oblongs, described in the margins as "12 chains and 28 links in length and 5 chains and

60 links in breadth including the alley which is 16 links wide," which suggests that speed required the original plan to be modified. Putnam's plan imaginatively incorporated some of the Hopewell mounds into the town's squares, but in those first years the settlement was plain rather than beautiful, and the rows of log cabins had the look of a military barracks.

In the summer of 1788 Putnam sent out several survey parties under the leadership of men like Israel Ludlow and Absalom Martin, both of whom had worked on the Seven Ranges. He drove them hard, paying a niggardly two dollars a mile despite the rough terrain and, unlike Hutchins, expecting a high daily mileage to be completed. They were required to survey not only the exterior lines of the townships but to mark each mile with a post or a blazed tree so that it could serve as a corner of the sections. As on the public survey, notes had to be taken of any salt licks, streams, or useful features. But Putnam was not unrealistic. Congress had dropped the requirement that the meridians run due north-south, and he did not fuss about it either. Fixing due north required astral or sun sights, and correcting for local magnetic variations. It all took time. Magnetic north wandered slightly but was close enough. Besides, as a surveyor himself, he knew the difficulties of keeping a straight line in wooded terrain. "In a covered Country," he wrote a little defensively, "altho great pains is taken, we must be very fortunate if we don't fall into many errors."

Israel Ludlow was not very fortunate. His north-south meridians were wayward, his east-west parallels haphazard, and his grid prompted one modern surveyor to observe that "scarcely two sections could be found of the same shape or of equal contents." On the other hand, he worked quickly. One year when Putnam gave comparable assignments to his teams, he noted that Ludlow had completed his before Martin was two-thirds finished, and a third surveyor named Biggs had barely reached halfway. And that was in land that defeated others altogether. "Before going a mile," C. H. van Orden, a fellow surveyor, wrote of the hilly forests in eastern Ohio, "I discovered it was impossible to do accurate chaining in such a broken country, where the hills were so steep it was often with difficulty they could be climbed."

The surveyors used only the circumferentor, for direction, and Gunter's chain, for distance. They planted posts or blazed trees at the corners, and, in

that rough country where one milepost might be hidden from the next by a rise in the ground, they blazed other trees to give a line of sight. In every sense of the phrase, they were blazing a trail. In their notebooks would be jotted hasty indications of the line markers' locations: "From the Mile post a Spanish Oak 14 inchs diameter bears N[orth] 83°15' E[ast] 30 [chains] ½ links distant with 6 notches and a blaze under them—and a white Oak 20 inches diameter bears S20°30'E 50 links distant marked in the same manner."

What Putnam valued in Ludlow was that he got the job done. When no one else had even set foot in the country, it was as much as anyone could ask to have a plat, a note of the licks, and the springs, and the rivers, and a description of the land detailed enough to make a sale to a customer back east. John Cleves Symmes liked that, too, and in 1788 contracted Israel Ludlow to survey most of the Miami Purchase.

A New Jersey judge, Symmes had persuaded the Continental Congress in 1787 to sell him one million acres of the Northwestern Territory between the Great and Little Miami Rivers, for a down payment of just $82,000 in paper money and military warrants. His deal had the backing of Jonathan Dayton, who was shortly to become Speaker in the new Congress, and in return Dayton was able to secure the best part of four townships, some of which became the present city named for him.

Clever though he was as a deal maker, Symmes made the mistake of letting Israel Ludlow work unsupervised, and the short history of the Miami Purchase was punctuated by violent disputes between settlers quarreling over Ludlow's boundaries. Even when he had Putnam breathing down his neck on the Ohio Company tract, there had been a drunken look about Ludlow's squares; left to himself on the Miami Purchase, they became chaotic. Worse still, he managed to survey and so make available for sale thousands of acres that lay outside the Purchase, including the site of what is now Dayton, Ohio. His mistake meant that those who bought the land from Symmes were left with no clear title to their property and after protracted legal argument had to pay for it again, this time to the United States. Among the victims were not only Dayton but General Arthur St. Clair, the first governor of the Northwestern Territory, and, happily enough, Israel Ludlow.

Although settlers began to arrive almost from the moment that Marietta was founded, Putnam had to take time out in 1789 to persuade Congress to defer payment of the next installment on the price of the company's land. It was not a job for which he was suited, and for probably the first time in his life his courage failed. A frantic letter arrived in Ipswich, Massachusetts, addressed to the Reverend Cutler. "Your presence here is very much wanted," Putnam wrote. "I have not time to explain to you the necessity of your being on the spot as soon as possible, can only tell you that if you have any regard to your own interest, you will set out for New York without loss of time—I repeat it my dear friend come on without loss of time—Come my dear friend don't fail." The silver-tongued cleric raced to Putnam's assistance. With the public survey suspended, Congress had no real option but to keep faith with the company, and Cutler was able to win a temporary suspension of payments.

Having survived the crisis, the new settlement started to flourish. By the time Putnam returned in 1790, the first law court had been established at Marietta, the governor, Arthur St. Clair, was installed, and a school had been established. In November Putnam was joined by Persis and their family, numbering eight children and two grandchildren, bringing with them some of their elegant Massachusetts furniture, a set of spindle-backed chairs, a four-poster bed, a fine desk, and Persis's cello. Civilization had arrived in the forest.

That year President George Washington paid the new settlement a glowing compliment. "No colony in America was ever settled under such favourable auspices as that which was first commenced at the Muskingum," he announced. "I know many of the settlers personally, and there never were men better calculated to promote the welfare of such a community."

The one Washington knew best was the settlers' leader, and in his autobiography Putnam admitted modestly that whatever troubles the world might have heaped upon his straight back, he had always found his commander in chief's opinion of him to be "no small sorce of consolation."

SEVEN

The French Dimension

*I*N THE AUTUMN of 1784, Putnam's "Arch Enemy" arrived in Paris as the minister of the new United States to France. Thomas Jefferson's entourage included his daughters, slaves, cook, secretary, and valet, so his first duty was to find a good-sized residence to serve as both home and embassy. He eventually settled on the Hôtel Langeac, a building in a new development between the Champs-Elysées and the rue Saint-Honoré.

Five years earlier the site had been a nursery garden belonging to the royal family, but the king's brother, the comte d'Artois, had decided it was more valuable as real estate, and had it measured, divided into lots, and sold off for housing. In doing this, he made himself simply the most prominent of more than a dozen dukes, marquises, and counts who had joined forces with architects and developers to create a real-estate boom that was transforming the French capital. Every day new buildings were going up on what had been the market gardens that fed Paris, pushing the city limits out toward Montmartre and the old parade ground of the Champ-de-Mars.

What was happening in Paris could also be found in other more developed parts of France like Lyons and Bordeaux. In the course of the eighteenth century, more than 150 years after England, and almost a century after the American colonies, France had started to discover the rapturous delight of property. Common land was enclosed, woodland fenced off, grazing rights taken into private hands, and aristocratic estates sold to wealthy merchants. As always, the surveyors were the litmus paper that indicated the changeover from feudal to private ownership.

In Lorraine in eastern France, for example, an accelerating program of enclosures was putting an end to the old open-field system of farming, and tracts of land were being engrossed into single fields. Instead of the old peasants' measures based on the amount of ground needed to feed a family, the *géomètres*, or surveyors, used a precisely defined *perche*, consisting of 18 *pieds*, each of which measured exactly 12.789 inches. One hundred square *perches* made an *arpent*, and the area could all be drawn to scale on a map and defined as the absolute property of a single owner.

In Saint-Etienne, near Lyons, the surveyors were not so far advanced, but even there they boasted, "We have put all these defective [measures] in good order, so that in each district their content is regulated in either *perches*, *pas* [64 inches] or *pieds*." Striking a cautious note, they advised that it was best to consult the peasants before enclosing open fields, but the fact that the advice was needed suggests that the contrary often took place. The sort of "defective" measures they were replacing were those that the peasants who worked the land instinctively understood, and in 1789 an assembly of farm-workers meeting in Bourges in central France protested bitterly against the new, exact units. "The *arpent* is not divisible by *perches* or *pieds*," they insisted, "but by *journées* which means by fields that one man is able to plough in one day; according to local custom, one *arpent* is equal to sixteen *journées*." The *journée* was the quintessential variable unit, equivalent to the daywork in England, the *Morgenland* in northern Germany, the *giornata* in Italy, or the *journal* in Catalonia. Wherever it continued to exist, so did the feudal economy in some form, and a true land market could not fully develop until it was superseded by exact measures of length and area.

In his classic work on measuring titled *Mètrologie* published in 1780, the author, Alexi Paucton, emphasized the superiority of the new invariable measures over the old version: "They are the rule of justice and the guarantee of property which must be sacred." It would be another fifty years before Pierre-Joseph Proudhon riposted, "Property is theft," but by then it was too late—the land had been stolen, and property was everywhere.

The introduction of exact units into land measurement had been made possible by the work of France's center of scientific research, the Académie des Sciences. It had been established in 1668 with the primary task of providing "more accurate maps than we have hitherto possessed," and since then one single project had dominated the Académie's activities: the mapping and exact measurement of France, a project that had forced its scientists to measure the world. Expeditions had been sent to take exact star sightings on the equator in Peru and beyond the Arctic Circle in Lapland, and for more than a century various teams had been triangulating the length of France from the English Channel to the Mediterranean, as well as every inch of its coastline. The names of those involved—Jean Picard the astronomer, and three generations of the Cassini family, including the greatest, Jacques Cassini—were famous not only in their own country but, as Jefferson himself had proved, across the Atlantic.

To carry out its work, the Académie needed a precisely defined and unchanging unit of measurement, and in 1766 it announced that France's basic measure, the *toise*, had been standardized at 76.734 inches in length. At the same time, an exact definition of the *livre*, or pound, was established based on weights dating from Charlemagne's reign, and standard copies of both these scientifically approved measures were sent to all eighty provinces in France. In the marketplace where the local magnates jealously guarded their privileges, the Académie's standards did little more than add a fresh possibility to the confusion of existing measures. "The unending proliferation of measures is quite beyond imagination," wrote Arthur Young, an agricultural expert who visited France in 1789. "They differ not only in every province, but in every canton too, and almost in every town; the differences drive people to despair." To the *géomètres*, however, the new, uniform *toise* was what made their work possible.

As in Tudor England, in eighteenth-century France it was those in control of the measures who were able to establish their rights of ownership, while the others who had only feudal custom to justify their occupation of the land were dispossessed and forced to take to the roads looking for casual labor. In the early days of the French Revolution, when the whole country was in a state of tension, the appearance of crowds of wandering beggars would frequently trigger rumors of invading armies.

Convinced as he was that working the land was the guarantee of democracy, Jefferson found it intolerable that the dispossessed had nothing while the wealthy few had so much they could afford to keep huge estates as uncultivated game parks. In the fall of 1785, as he was walking in the countryside outside Paris, he encountered one of these landless poor, a widow struggling to bring up two children on the eight cents a day she could earn as a laborer, and later that day Jefferson wrote furiously to James Madison, "Whenever there are in any country uncultivated lands and unemployed poor, it is clear that the laws of property have been so far extended as to violate natural right. The earth is given as a common stock for man to labor and live on."

The injustice of it convinced him of the necessity for political reform in France, but it also confirmed his deep dislike of those speculators who hoarded huge swaths of American land, keeping it uncultivated until the price rose and keeping out the poor farmers who would have worked the soil. The entire nation suffered from their greed because, as he told Madison, "the small landholders are the most precious part of a state." Without stringent precautions, he felt, the Western Territory might produce a second France rather than a new republic.

Money was required to make the most of the investment in property. While some modernizing aristocrats, like Jefferson's landlord, the comte d'Artois, took advantage of the opportunity, the prime beneficiaries were merchants, financiers, and industrialists. To this class the need for uniform measures was obvious. It would have been impossible for a manufacturer like Jean-Baptiste de Gribeauval, whose armaments factories were to produce Napoleon's incomparable artillery, to cast thousands of cannon of the same caliber without unchanging and exact units of measurement.

Yet even conservative members of the aristocracy were adapting to the changes taking place. Improvements in agriculture were producing larger harvests, and in cash-poor economies like western and southern France, landowners took their share of the growers' increasing profit by arbitrarily increasing the size of measures in which corn-rent and other dues were paid. That was the way feudal economics had always worked. The measures varied because in theory the amount of money was fixed—and in practice there were few coins in circulation. To see it the other way around required a paradigm shift in thinking from ancient to modern.

In the summer of 1788, as Rufus Putnam and his party were settling into Marietta, a disastrous harvest brought to the surface all the smoldering resentment against the new exact measures and the old expanding measures. With France approaching the brink of bankruptcy, King Louis XVI was forced to call a meeting, for the first time since 1614, of the nation's ancient representative body, the Estates-General, and the peasants took the opportunity to voice their grievances in catalogs drawn up for their delegates called *cahiers de doléances*.

The call for justice was deafening. Across Brittany in the northwest the complaints were almost identical. "The measure at Plumaudan near Dinan has grown much larger," declared one community, and close by in Plancoët another protested, "The measure they apply to the grain is arbitrary," while from the port of Quimper the assembly insisted, "The nobles' measure grows larger year by year." "We demand," wrote a commune in northeastern France, "that the lords to whom we are obliged to pay corn tax be obliged to keep standards at the manor house, properly attested according to the old measure."

References to "the old measure" appear repeatedly, indicating how widespread was the anxiety at the changes being introduced. Most *cahiers* assumed there had once been a single, uniform system, and that was what they wanted to get back to. As an assembly in central France put it, echoing the ancient policy of France's kings, "Let there be in the Kingdom only one god, one King, and one law, one weight and only one measure." Contemplating that lost simplicity, a commune south of Paris voiced its frustration in a furious question that was implicit in every *cahier*: "Why should the seigneurs and the privileged clergy enjoy the right to please themselves as to what measure of grain is binding on their estate?"

Jefferson's most frequent guest at the Hôtel Langeac was convinced he had the answer to their complaints. Dismayed by the seeming impossibility of establishing a uniform system on France's chaotic weights and measures, the marquis de Condorcet, permanent secretary to the Académie, believed that the science and technology developed by the Académie made it possible to devise a system of measurement that would remove all the injustices occasioned by the old arbitrary units. For the first time since human society began, measurement could be based on units derived from scientific inquiry rather than parts of the body or human activity.

Condorcet was not only a leading mathematician and philosopher but a passionate modernizer and idealistic advocate of Enlightenment values. His wife, Sophie, shared his radical ideals, and since their only child was born in April 1790, Paris gossip would take pleasure in assuming that she must have been conceived on the most erotic night in radical history—July 14, 1789, when the Bastille was stormed.

Condorcet knew Jefferson's reputation as the author of the Declaration of Independence, and with characteristic enthusiasm immediately sought him out when Jefferson arrived in Paris. Not only did he share Jefferson's rationalism; they were the same age and resembled one another in their tallness, their physical awkwardness, their enthusiasm for science, and, most of all, their divided temperaments. "Between the sharpness of his intellect and the goodness of his heart," observed Amélia Suard of Condorcet, "there was always a contrast that I found singularly striking." The description could have served equally well for the American minister. But where Jefferson shambled and talked of his concepts in easy discursive style, Condorcet was tense, white-faced, nervously chewing his nails, and expressing his thoughts with wound-up fury—"a snow-capped volcano" was one friend's unkind description.

During the five years that Jefferson spent in Paris, he and Condorcet fed off each other's ideas. Condorcet wrote a pamphlet in praise of the American Revolution, while Jefferson translated Condorcet's essay against slavery into English. And to both, it became obvious that science and democracy were natural allies. "A good law ought to be good for all men," according to one of Condorcet's maxims, "as a true [scientific] proposition is true for all men." Finding a scientific basis for measurement would remove control of weights and measures from the privileged and powerful, and make it available to anyone who knew arithmetic or could consult a scientist. In words that Jefferson would echo in the United States, Condorcet asserted that "the uniformity of weights and measures cannot displease anyone but those lawyers

Marquis de Condorcet

who fear a diminution in the number of trials, and those merchants who fear anything that renders the operations of commerce easy and simple." Nor did they disagree about the scientific basis for a new, uniform system of measures.

In "Some Thoughts on a Coinage" Jefferson had chosen the length of the equator, but Condorcet had already investigated the question for the French government. The French earth measurers had proved that the globe was flattened at both ends like an orange, rather than forming a perfect sphere, which created almost insuperable difficulties in measuring its circumference accurately. Condorcet's report in 1775 concluded that the length of the second's pendulum, slightly more than 39 inches, ought to be the basic unit. Within a few months of meeting Condorcet, Jefferson had not only changed his mind but convinced James Madison, with whom he remained in close touch throughout his years in Paris, of the pendulum's merits. Madison in turn wrote to James Monroe in 1785, urging that it would be "highly expedient, as well as honourable to the federal administration, to pursue the hint which had been suggested by ingenious and philosophical men, to wit: that the standard of measure should be first fixed by the length of a pendulum vibrating seconds at the Equator or any given latitude."

Jefferson always remembered his time in Paris with affection. He had been sent there to sign treaties of friendship, to negotiate loans, and secure trade advantages for the United States, but what he enjoyed above all was the opportunity to discuss the ideas of the Enlightenment with others of like mind. In his fragment of autobiography, he wrote affectionately but in general terms of friends like Condorcet, remembering especially "their eminence in science, the communicative dispositions of their scientific men, the politeness of the general manners, the ease and vivacity of their conversation." But Condorcet paid a more specific tribute to their intimacy. In 1794, as he was about to be arrested by Maximilian Robespierre's Jacobins and knowing he faced certain death, he drew up a will entrusting to two friends the education of his daughter. One of those friends was Thomas Jefferson.

*I*N MARCH 1786 Jefferson paid a short visit to Britain and returned to Paris with all his prejudices against the former colonial power reinforced. "That nation hates us, their ministers hate us, and their King more than all other men," he wrote to a friend. He could find only two things in its favor: the beauty of its gardens—"in which it surpasses all the earth"—and the excellence of its technology. "The mechanical arts in London are carried to a wonderful perfection," he admitted, and nowhere was that perfection more apparent than in the making of measuring instruments. Throughout the eighteenth century, professional scientists and wealthy amateurs interesed in research had encouraged manufacturers to create ever more accurate chronometers, micrometers, scales, and gauges, whatever helped to make measurement more precise. John Dollond's microscopes, John Bird's telescopes, and George Adams's theodolites were famous across the Western world. A few years later, although heavily in debt, Jefferson was prepared to pay more than $200 for an advanced theodolite, or "Universal Equatorial Instrument," made by London's leading instrument maker, Jesse Ramsden.

There was a paradox here. In the centuries since Queen Elizabeth created the framework of the country's measuring system, official weights and measures in Britain were so badly made that no accurate standards existed except for the 1601 yard. In 1758 a parliamentary committee under Lord Carysfort investigating the state of the country's measures discovered the problem began with the legislation, which was so contradictory that one act alone managed to define the capacity of the Winchester bushel, used on both sides of the Atlantic, in four ways. Compounding the confusion, the master standards held by the government were so clumsily constructed that the committee was unable to find a single bushel container or a single weight in the office of the Exchequer, or the Tower of London, or the Mint, that was the same size as any of the others. And the copies made from these standards for use in cities throughout Britain were worse still.

"The Unskilfulness of the Artificers [makers], joined to the Ignorance of those who were to size and check the Weights and Measures in use, occasioned a great Number of different standards to be dispersed throughout the kingdom which were all deemed legal yet disagreed [with each other]," the committee reported. Poorly made copies of the bushel and yard were sent out to cities and "became themselves again the Standards for other Copies, made and used in different [places] by Artificers yet less skilful than the first Makers. . . . Thus every Error was multiplied till the Variety . . . rendered it difficult to know what was the Standard and impossible to apply any adequate remedy."

Despite this damning report, the British government did nothing. It was left to the Royal Society, the center of scientific and technological research, founded in 1663, to try to fill the gap. Unlike the Académie, the Royal Society was not government-funded, but such was its eminence that scientists in both Britain and America looked to it for leadership, and in 1743 Benjamin Franklin took it as a model when he founded the American Philosophical Society. In 1742 it had had a yard standard built to exact specifications based on Elizabeth's brass yardstick, and on its advice the Carysfort committee commissioned an even finer example from John Bird.

His yard, made in 1760, was a work of art, a 39-inch brass bar with the two hairline grooves exactly 36 inches apart—as gauged against the Exchequer yardstick of 1601—set into gold studs in order to avoid distorting the bar itself. As a secondary check in case it was destroyed, Bird's bar was measured against a second's pendulum oscillating in a vacuum. Eventually, it would become the master yardstick against which the standards of Britain and its empire would be measured, and when yet another London instrument maker, Edward Troughton, came to manufacture the measure that the United States ultimately adopted as its basic unit of length, Bird's beautifully made yard served as the final check on its accuracy.

Yet there was still little pressure on the British government to standardize measures of weight and capacity. Local communities maintained their own scales and containers, just as they kept their own time. In the nineteenth century the arrival of the railroad with its unvarying timetable forced formerly isolated towns and villages to harmonize their clocks with their neighbors'. In similar fashion, when the spread of industrial standards and better roads brought in the wider world with its own ideas of what a pound should weigh or a bushel should contain, pressure grew for uniform measures.

One of the strongest influences toward uniformity came from the inventor James Watt, who in 1774 left Glasgow in Scotland to join Matthew Boulton in the English Midlands to manufacture a more efficient steam engine. To measure the increased efficiency that came from adding a condenser to the existing model, Watt invented a new unit, the horsepower, which he defined as the amount of energy necessary to raise 33,000 pounds one foot in one minute. Wherever Watt's improved steam engine was used—and it was to become the workhorse of the Industrial Revolution, providing the power to perform tasks that once required hundreds of people—its efficiency could only be measured with a pound weight recognized by the manufacturers. Even Jefferson, who thought little better of manufacturing than he did of George III, was astonished by the results when he visited a steam-powered flour mill during his English visit. "In the arts the most striking thing I saw there, new, was the application of the principle of the steam-engine to grist mills," he wrote. "I saw 8 p[ai]r. of [grinding-]stones which are worked by steam, and they are to set up 30 pair in the same house."

Pressure for standardizing came from a different direction when the French passion for the exact measurement reached across the English Channel. In 1784, as the triangulation of France was about to reach the northern coast, Cassini de Thury, the third of his family to be in charge of the country's measurement, suggested to the Royal Society in London that the triangles should be continued across the Channel. There had already been cooperation between the Académie and the Royal Society—a copy of the 1742 yard had been sent to Paris, and the length of the *toise* communicated to London—and de Thury's suggestion would enable the distance between the observatories in Paris and Greenwich to be precisely measured so that their observations could be coordinated.

The Royal Society lacked the resources for the sort of strategic approach that the Académie had taken to geodesy. Consequently in Britain, while numerous estate and county maps existed, and the Highlands of Scotland had been systematically surveyed for military reasons following the Jacobite rebellion in 1745, there was nothing like the national map based on exact triangulated measurement that France by then possessed. However, General William Roy, the formidable Scots disciplinarian who had masterminded the Highland survey, was determined to create a British map to rival the French, and he almost single-handedly persuaded the Royal Society to back the Académie's proposal. The one essential quality of all great surveyors is an infinite desire for greater accuracy, and Roy passed that test easily.

To calculate their baselines, the French had used wooden measuring rods that were protected in boxes and checked daily against an iron bar exactly one *toise* in length. These were tried but discarded by Roy because even inside their boxes the wood tended to expand slightly in a damp atmosphere. He replaced them with calibrated, 18-foot-long, glass rods, manufactured by the great instrument maker Jesse Ramsden. They needed to be carried in special containers and supported on trestles to prevent them from fracturing, and they were checked against the Royal Society's 1742 yard, which had been marked out with a beam compass fine enough to show subdivisions of one eight-hundredth part of an inch.

However, nothing shows Roy's fanaticism to better advantage than his choice of a theodolite. Jean-Charles de Borda, the French cartographer responsible for the final stage of triangulation, had invented a beautiful instrument that carried observation to a new order of accuracy. It had long been recognized that one sight on a distant mark through a theodolite's narrow focus might contain an error, and geodesists as a matter of habit would take five or six observations and average the results. Borda's genius was to build into an instrument weighing no more than 20 pounds a mechanism that automatically averaged ten observations, and thereby reduced the margin of error from fifteen seconds' deviation to little more than one second.

Borda's repeating theodolite was the pride of French geodesy, but when it was offered to Roy, he turned it down. It had one major drawback—for maximum accuracy the mechanism had to be adjusted constantly—but that was not Roy's objection; he wanted more than its maximum. For that he went back to Jesse Ramsden and commissioned one of the most precise instruments that had ever been constructed in surveying history. The device was a monster, a mass of gleaming brass weighing more than 200 pounds with a 3-foot-diameter circle, and an achievable accuracy on its giant compass of less than one second. To move it, Roy needed a company of redcoats on hand at all times, but it offered measurements as exact as the eighteenth-century could aspire to. While he lived, it soothed the search for perfection in the general's Calvinist soul.

Such exquisite engineering takes time, and it was five years before the line of triangles reached from Greenwich to Dover, from where sightings across the 22-mile stretch of the English Channel connected the British survey to the French. The accuracy of the work, however, could not be questioned; triangulation showed that the distance between the observatories in Greenwich and Paris, approximately 163 miles, was just 15 inches shorter than that calculated by astronomical observation. Roy, unfortunately, had little time left to apply his ruthless zeal to the larger task of measuring Britain before he died in 1790, but his appetite for accuracy brought into being the Ordnance Survey, which was to create a geodetic map of Britain and led to triangulated maps being made of almost every square mile of the British Empire from the immensity of India to the compactness of Cyprus.

A theodolite.

In April 1790, while this example of international cooperation was still in progress, the bishop of Autun, Charles Maurice de Talleyrand-Périgord, sent a letter to a British parliamentarian, named Sir John Riggs Miller, proposing a still-closer alliance between the two countries. "I understand that you have submitted for the consideration of the British Parliament, a valuable plan for the equalization of measures," wrote Talleyrand. "I have felt it my duty to make a like proposition to our National As-

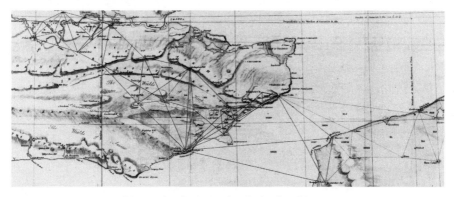

Survey triangles connecting England and France.

sembly. It appears to me worthy of the present epoch that the two Nations should unite in their endeavour to establish an invariable measure and that they should address themselves to Nature for this important discovery. Too long have Great Britain and France been at variance with each other, for empty honour or for guilty interests. It is time that two free Nations should unite their exertions for the promotion of a discovery that must be useful to mankind."

Riggs Miller was a bore—the evidence lies in his interminably windy speeches—but he was also persistent and driven by genuine outrage at the inefficiency and injustices caused by the state of British measures. In 1789 he had started a campaign to have them improved, and night after night he kept the House of Commons sitting late with his accounts of the scandalous discoveries made by the Carysfort committee, which, he declared, bred "abuses fertile in confusion to commerce, and in distress and difficulty to the poor." At Riggs Miller's request, the House of Commons ordered all market towns to report what measures they used, and in February 1790 when he addressed the House again, he had enough evidence to show that the situation had deteriorated since the Carysfort report. Because there were no fixed measures, market traders exploited their customers, and landlords their tenants—"the poor thresh out corn and other grain by the largest customary bushel and buy their bread by the smallest," declared Riggs Miller. "Nothing could be more degrading, more absurd, and more preposterous than that [measures] which should be the clearest and plainest . . . should prove to be undefined in law, obscure in practice and involved in perplexity."

The time had come to reform the entire confused system, he insisted, and introduce "one general standard of weights and measures to be observed throughout the kingdom." In April 1790, after receiving Talleyrand's letter, he initiated a debate in Parliament and proposed the adoption of a new scientifically based set of decimal measures. There were several possible bases that could be chosen, but his preference was for the length of a second's pendulum, and as a patriotic Briton, he specified in particular that "the pendulum which vibrates seconds at London, is the most proper standard for Great Britain, and a medium for all Europe."

To add strength to his argument, he read out the letter from Talleyrand, who, at the urging of Condorcet, had just submitted a proposal to France's National Assembly calling for a new system of weights and measures. It, too, was to be based upon the second's pendulum, but its length was to be measured naturally enough at the latitude of forty-five degrees, which happened to be where Paris was situated. "A turd in a silk stocking" was how Napoleon described the smoothly sinister Talleyrand, and probably any House of Commons M.P. still awake at the end of Riggs Miller's interminable speech would have yawned his agreement. Nevertheless, they agreed to appoint a parliamentary committee to examine the idea that Riggs Miller put forward.

What makes that month of April 1790 so extraordinary is that it marks the moment when, after 10,000 years of traditional measurement, the decision to look for a new, scientific basis was taken simultaneously in not two but three different capitals. At the very moment when London and Paris were embarking on the road to reform, Thomas Jefferson in New York was formulating a new system of weights and measures for the United States. And in all three cities, the reformers were considering the same solution—decimal units based on the second's pendulum—to the same age-old problem of local, variable measures. Such complete synchronicity might have moved mountains—even shifted public opinion.

EIGHT

Democratic Decimals

ONLY THE NEED TO SETTLE his teenage daughter in an American school took Thomas Jefferson away from the congenial company of French scientists in September 1789. He returned to find a United States with a new Constitution and Bill of Rights, a new Congress, and a new president who insisted that Jefferson take the post of secretary of state in the new federal government.

When George Washington addressed the Congress for the first time on January 2, 1790, he outlined with some care the most pressing matters that he thought it should consider for action. The first two were defense and the economy, and immediately after them came the need for a uniform system of weights and measures. Acting on this suggestion, the House of Representatives requested the new secretary of state on January 15 to draw up "a proper plan or plans for establishing uniformity in the Currency, Weights and Measures of the United States." Consequently, the first job waiting for Thomas Jefferson when he arrived in New York, the temporary capital, on April 15 to assume office was one for which he was particularly well suited, both by temperament and by his years in Paris and friendship with Condorcet.

Soon after Jefferson began work, Congress passed a bill to create "the Territory of the United States South of the River Ohio," which comprised lands formerly claimed by Virginia and South Carolina and were now the territory of Tennessee with some land farther south. A governor was appointed, boundaries were defined with the promise of a detailed survey to come, and the timetable of evolution to statehood, as proposed by Jefferson in 1784, was laid out. Already settlers were pouring into the region. Together with the spread of migrants into the Northwestern Territory and the newly created states of Vermont and Kentucky, this expansion of the United States underlined the urgency of Jefferson's work.

The settlers brought with them a gallimaufry of measures: Rhineland *Ruthin* (rods) and Scots miles and Irish acres for measuring land; Scots mutchkins for beer; English flitches for bacon, fardels for cloth, and hattocks for grain; German *Globens* for flax, *Quentchens* for gold, and *Zehnlingen* for lambskins—in all an estimated 100,000 different units. For genuine frontier families who lived off what they raised, the variety of these local units presented no problem. Among themselves they used their own lengths and capacities. In western Pennsylvania these tended to be Rhineland measures, in Vincennes on the Wabash River they were French, in New York, Dutch. The only goods that needed measuring by strangers were the ones they had to buy—flour, coffee, and sugar at the market, and cloth from peddlers—and no matter what their nationality, country dwellers knew that when they came to town, the scales were literally loaded in favor of the merchants who owned them.

The problem had begun with the arrival of the original colonists. Virginia made its initial attempt to legislate against fraudulent measures within a generation of the first plantation, a second in 1646 when legislators accused both Dutch and English merchants of practicing "deceit and diversity of weights and measures," came back for a third try in 1661 with an act whose preamble began, "Whereas dayly experience sheweth that much fraud and deceit is practised in this colony by false weights and measures . . . ," and followed this up with more legislation in 1734—the fourth attempt in a century, accompanied by the sad explanation, "Forasmuch as the buying and selling by false weights and measures is of late much practised in this colony to the great injury of the people."

Virginia was not unique. The phrase "frauds and deceits" appeared so often in colonial legislation that it became a cliché. It was used in the preamble to a Maryland law of 1671 referring to "much fraud and deceit [that has been] practised in the province by use of false weights and measures." An eighteenth-century North Carolina law used it in the plural—"Whereas many notorious frauds and deceits are daily committed by false weights and measures . . ."—as did the New Jersey Act of 1725 calling for "one just weight and balance, one true and perfect standard for measure, for want thereof experience has shown that many frauds and deceits have happened." In New York, where the range of English measures was complicated by the continuing existence of Dutch units, and by the tendency of unscrupulous retailers to buy cloth by the English ell, which was almost a yard, but sell by the Flemish ell, barely half its length, the law plaintively pointed to the need for "one true and perfect standard and assize of measure among them; for want whereof experience shews that many frauds and deceits happen, which usually fall heavy upon the meanest and most indigent sort of people, who are least able to bear the same."

For all its democratic assemblies, its fine Constitution, and its splendid Bill of Rights, down in the marketplace the new United States often looked remarkably like the old colonies. The legal bushel in Pennsylvania, for example, was supposed to be wide and shallow, so that the amount of grain heaped above the rim would equal one-third of the amount in the container itself. In Philadelphia, however, shopkeepers made the bushels deep and narrow, so that when filled they would take only a small heap above the rim.

In Maryland, farmers tried to take court action against grain merchants who combined to introduce a larger half-bushel measure for use in buying grain but paid the same price. However, the justices of the county courts consisted of urban citizens in good standing, like shopkeepers and grain merchants, and they were also responsible for weights and measures. And so, as a contemporary critic, John Beale Bordley, noted, the merchants simply appointed sympathetic justices whose job was to "order new standards, which had great weight in quieting opposition to the new half bushel measure."

It would be hard to imagine a problem more attractive to Thomas Jefferson's superlative combination of lucid intelligence and fierce emotion. The typical victims of false measures were those small farmers on whom the health of democratic government depended, in his opinion, but who were easy prey when they came to market to sell their produce and stock up on necessities like cloth and sugar. The abuse of those Jefferson most admired by the urban traders he most disliked was an affront to what he termed "the democratic principle," and was thus intolerable. But because control of weights and measures was a source of profit, it would always end up in the hands of the powerful. The only sure recourse was to make the system simple, uniform, and so transparent that "the whole mass of the people . . . would thereby be enabled to compute for themselves whatever they should have occasion to buy, to sell, or to measure, which the present complicated and difficult ratios place beyond their computation." Democracy meant decimals.

Almost inevitably, given their ideological differences, opposition to reform came from financiers and merchants like Jefferson's old nemesis, Robert Morris, who refused to believe that change was necessary. The conflict had moved on in the six years since the defeat of Morris's bid for a new currency and, closer to his speculative heart, the chance of running the U.S. Mint. In 1784, while Jefferson was in Paris, Morris had negotiated, behind the American minister's back, an agreement with the French tax-collecting authority, the Farmers-General, that he alone would supply American tobacco to France. Two years later, Jefferson hit back by persuading the French government to break Morris's monopoly and buy tobacco from other American, and especially Virginia, suppliers. In 1790 both men were in New York, Morris riding a rising tide of land speculation and a soaring market in government securities, and Jefferson determined to give republican farmers an even chance against the chicanery of city sharks, shopkeepers, grain merchants, and land speculators.

The task he had been given by Congress appealed to him, therefore, on every level, not least because it took him back to the mathematical problems with which William Small had wooed him into Enlightenment thinking. His only doubt, as he admitted in a letter to the astronomer David Rittenhouse, was whether he could do justice to such a daunting challenge: "Five and twenty years ago I should have undertaken such a task with pleasure because the sciences on which it rests were then familiar to my mind and the delight of it. But taken from them . . . I have grown rusty in my former duties."

His chief handicap was a lack of reference books—he had expected to return to Paris, and most of his library was still there, with the remainder in Monticello. There was also New York itself; the secretary of state hated the city. It represented everything he most disliked about urban life, being full of noise, odors, and people on the make. During the Revolution, the British had bombarded and occupied it, the Americans had thrown up barricades, fires had broken out, and buildings had collapsed. By the time peace came, weeds grew between Broadway's cobbles; but nothing could alter the magnificence of New York's harbor, and at once an orgy of rebuilding began. Traders

Thomas Jefferson

returned, and they were accompanied by the financiers and shipbuilders who made their business possible. The presence of Congress meeting in City Hall on Wall Street brought in printers and publishers, and gave trade to coffee shops, boardinghouses, and restaurants.

When Jefferson arrived in April 1790, the reborn New York was overflowing its pre-Revolution boundaries at the foot of Manhattan. Jefferson's house in Maiden Lane in Lower Manhattan, several blocks north of Wall Street, should have been on the outskirts of the city; instead it had been engulfed by the thunder of construction as new red-brick houses were built for the expanding population. New York, said Jefferson with the moderation to be expected of a rational man, "is a cloacina [sewer] of all the depravities of human nature." Not only was he shut up in the cloacina as spring merged into the oven-heat of summer, he had set himself a Herculean task: to invent from scratch an entire system of measures.

He had, however, one invaluable colleague in James Madison, for whom reform had become no less of a crusade. Among Madison's many quiet qualities was a tireless capacity for research, and he found a pamphlet by Robert Leslie, a Philadelphia watchmaker, who recommended a solid rod for measuring seconds. The advantage of using a rod instead of a weighty bob hanging from a wire was that it allowed the length to be calculated with greater precision. Jefferson adopted the idea without hesitation.

Jefferson's conversations with Condorcet and other French scientists were still fresh in his mind, and he must have had available his notes, "Some Thoughts on a Coinage," because with astonishing speed he was able to sketch out the principles of the new system and its major divisions by the middle of May. Then, after three weeks of intense and fiddly computation carried out in the noxious surroundings of New York, he was struck down by an intense bout of migraine. Years later he remembered it as "a severe attack of periodical head ach [*sic*] which came on every day at Sunrise, and never left me till sunset." At its worst, it forced him to sit in a room too dark for writing and to work out problems by mental arithmetic. "What had been ruminated in the day under a paroxysm of the most excruciating pain was committed to paper by candlelight, and the calculations were made."

Gradually, the severity of the attacks wore off, and Jefferson returned to daytime work. Less than two months after he received notice of Congress's request, he had completed the task. It was, and remains, a formidable feat of intellectual application.

He began by choosing the swing of Leslie's rod for the base unit: "To obtain uniformity in measures, weights and coins, it is necessary to find some measure of invariable length with which, as a standard, they might be compared." The size of the earth might have served, but Condorcet had convinced Jefferson it was impossible to measure the equator or the whole of a meridian accurately. On the other hand, Jefferson pointed out, "the motion of the earth round it's [*sic*] axis is uniform and invariable." The time that elapsed in the course of one revolution was everywhere the same and could be divided into 86,400 parts, each of which was known as a second. The length of a pendulum, or better still a rod, which took one of those seconds to swing from one end of its arc to the other, could also be established. "Let the Standard of measure then be an uniform cylindrical rod of iron, of such length as in lat. 45° in the level of the ocean, and in a cellar or other place, the temperature of which does not vary thro' the year, shall perform it's vibrations, in small and equal arcs, in one second of mean time."

In his first draft, Jefferson chose thirty-eight degrees north, the median line of latitude running through the United States, as the point where the pendulum or rod should be measured. But once he heard of the Condorcet-inspired reform that Talleyrand presented to the National Assembly, he changed it to forty-five degrees, the latitude of Paris, to harmonize with the French proposals.

Based on the second's rod, Jefferson then offered two solutions under the pretext that he was not clear how much change Congress wanted. Did they want a root-and-branch reform, similar to his decimal dollar system, in place of the old pounds, shillings, and pence? "The facility which this would introduce into the vulgar arithmetic," he explained, "would unquestionably be soon and sensibly felt by the whole mass of the people who would thereby be enabled to compute for themselves whatever they should have occasion to buy, to sell, or to measure, which the present complicated and difficult ratios place beyond their computation for the most part."

His own writings made it clear that this was what he himself wanted. But he recognized that it might be "the opinion of the representatives that the difficulty of changing the established habits of a whole nation opposes an insuperable bar to this improvement." His first task, therefore, was to convince them of the utter necessity for some measure of reform.

Because the first settlers came mostly from England, the weights and measures generally used in the United States were English, and, Jefferson explained with barely concealed disdain, "we must resort to that country for information of what they are or ought to be." Turning to the findings of the Carysfort committee, he asserted that Elizabeth I's measures of length were "brass rods, very coarsely made, their divisions not exact, and the rods bent." He accepted that in 1742 the Royal Society had had a new and very accurate yard standard made, but Americans could not know exactly how long it was because "they furnish no means, to persons at a distance, of knowing what this standard is."

This was not wholly honest—Elizabeth's yard, despite its construction, was astonishingly accurate, and it would have been easy to commission an exact copy of the Royal Society standard—but it served Jefferson's purpose to cut American measures free of the English standards. If units of distance like the foot and inch, and units of area like the acre, were based on the length of the second's rod, the existing units could be left unchanged, but they would be based on American science.

Then Jefferson turned to measures of capacity, and the sly turn of his humor can still be caught in the deadpan way in which he lists their absurdities:

> 8. *gallons make a measure called a firkin in liquid substances, and a bushel, dry;*
>
> 2 *firkins or bushels make a measure called a rundlet or kilderkin, liquid, and a strike, dry;*
>
> 2 *kilderkins or strikes make a measure called a barrel, liquid, and a coom, dry, this last term being antient and little used;*
>
> 2 *barrels or cooms make a measure called a hogshead, liquid, or a quarter, dry.*

Having listed every last one of them, he delivered the punch line: "But no one of these measures is of a determinate capacity." The contradictions in English law meant that each of them could be defined in at least eight and as many as fourteen ways.

Having demonstrated the measures' unreliability, he discarded them all, beginning with the wine gallon, because it was only important to "the mercantile and the wealthy, the least numerous part of society." He proposed in their place a new gallon of 270 cubic inches, which was roughly halfway between the most commonly used measures. All other capacities would be derived from it.

Then, with that scalpel clarity that enabled him to cut through constitutional and political complexity, Jefferson disposed of a problem that had caused complaints for centuries: "The measures to be made for use [shall be] four-sided, with rectangular sides and bottom. . . . Cylindrical measures have the advantage of superior strength, but square ones have the greater advantage of enabling every one who has a rule in his pocket to verify their contents by measuring them."

It was the simplicity of the square that made it democratic. As Jefferson explained, anyone could measure it. The same reasoning had led him to choose it for the survey of U.S. public land, because, in the words of the geographer William D. Pattison, "rectilinear land boundaries put it in the power of any settler, employing the most rudimentary means of measurement, to verify the contents of his purchase." But it is possible to guess that the matter went a little deeper in the Jeffersonian psyche.

Consciously or otherwise, he kept proposing the square as the solution to all sorts of problems. Administratively, he hoped that Virginia's counties would be square, so that they could be subdivided into square wards. "The wit of man," he declared, "cannot devise a more solid basis for a free, durable and well administered republic." Planning the nation's future capital city of Washington in 1792, he imagined it as a 10-mile-broad square, divided into smaller squares, whose functions he outlined in a letter to the president: "For the Capitol and offices, one square. For the market, one square. For the public walks, nine squares consolidated."

The square offered not just simplicity but symmetry, and when Jefferson came to make his imaginative constructions real, that is what chiefly emerges. Thus the groundplan of Monticello, the house he designed and spent forty years "pulling down and putting up," and where his heart always remained, reveals itself to be a central oblong, consisting of the hall and rotunda, which separates two balancing squares. The campus of the University of Virginia, the greatest pride of his later life, is a large square enclosing two oblong rows of pavilions and gardens symmetrically arranged on either side of a lawn. Even in unconscious details—like the *maison carrée*, the rectangular Roman building in Nîmes which Jefferson selected as the original for Virginia's capitol in Richmond, or the four-sided *partie quarrée* he remembered with such affection from his youth where he and William Small, George Wythe, and Francis Fauquier had set the world to rights—the square's perfectly balanced shape kept reappearing.

The balance in the square was also intrinsic to Jefferson's vision of democracy. As he made plain in a letter to James Madison from Paris in 1787, what he liked about the United States's proposed Constitution was the separation it made between the three elements of executive, legislature, and judiciary; but he insisted that it needed a fourth side to balance it, a Bill of Rights for the people. The shape ensured that forever afterward a debate would take place among the four parties—giving later generations the chance to hear a raucous echo of those Williamsburg discussions on which the young Jefferson had cut his intellectual teeth.

Thus the squareness of the containers was not just a detail in Jefferson's scheme of things. It was an integral part of the pattern that he wished to impose upon the whole structure of his ideal state. To rule out the smallest room for doubt, he specified the dimensions of each container and required its contents to be measured striked—that is, level with the rim—rather than heaped.

Jefferson's proposals for weights were also aimed at simplification. He wanted to amalgamate the two existing systems—the commonly used avoirdupois, whose range extended from drams to tons, and the specialist troy weights used mostly by jewelers and apothecaries. The basic unit of this unified system, which was essentially avoirdupois, would be the ounce, weighing exactly one-thousandth of a cubic foot of rainwater.

It was a wonderfully straightforward plan, but one designed almost entirely to set the House of Representatives thinking about the need for reform. Once it had succeeded in that object, Jefferson was ready to spring the more serious option on them. "But if it be thought," he continued, "that, either now, or at any future time, the citizens of the U.S. may be induced to undertake a thorough reformation of their whole system of measures, weights and coins . . . greater changes will be necessary."

In less than 1,000 words, he then outlined the first scientifically based, fully integrated, decimal system of weights and measures in the world. Its basic measure of length, derived from the second's rod, was a foot, which would be divided into 10 inches. A cube of rainwater, whose sides were 1 decimal inch long, was to weigh 1 decimal ounce, and 10 of these ounces would make 1 pound. The basic unit of capacity would be the bushel, which was to measure 1 cubic foot, that is to say, 1,000 cubic inches. Finally, the weight of the dollar was to be adjusted so that it came to exactly 1 decimal ounce.

The system was plain and elegant, and the manner in which the three dimensions of weight, length, and capacity were integrated, anticipating one of the great strengths of the metric system, gave it enormous coherence. None of its elements was original to Jefferson, save only his insistence that the bushel and other measures be "four-sided and the sides and bottom rectangular," but the synthesis of Newtonian physics with the democratic aim of "bringing the principal affairs of life within the arithmetic of every man who can multiply and divide plain numbers" was entirely his creation.

It was a remarkable moment in intellectual evolution. Measurement consists of abstracting one quality—length, mass, time, velocity—from an object or an event and giving it a numerical value. Originally, the unit used to make the measurement might have been personal—"your" foot; then it became social—"the" foot, an agreed average; now the foot had evolved into something scientific—a fraction of time.

The timing of Jefferson's report was critical. Any proposal for reform would encounter not only the outright hostility of traders but the passive resistance to change of any kind always felt by the majority of the population. There was one exception to this general rule. Out on the frontier, Germans, Scots, and Irish settlers, all reared with their own ideas of the distance of a mile, or the area of an acre, or the length of a rod, were quickly adapting to the basic reality of Gunter's 22-yard chain, to the 4,840-square-yard acre it measured out, and to the square-mile section that made up the 36-square-mile township.

English settlers who had moved to New Orleans or west of the Mississippi were equally adaptable. When it came to acquiring land, they were prepared to measure it as happily in French *arpents* and Spanish *labores* as in acres, and could learn to estimate a river frontage in *toises* and in *varas* without difficulty. By the best estimate, in 1795 more than one million acres, or almost 6,000 *labores*, in Spanish America were owned by families brought up on English land measures, and a stream of others were moving into the rice-growing, timber-producing, French-measured land around the Gulf of Mexico.

Jefferson's first attempt to link his new measures to the sale of public lands in 1784 had failed, but his instinct was sound. The future of decimalized measures was tied to that part of the United States where measurement was critical, and where there was a clear willingness to accept strange and novel units. That was a view accepted by the secretary of the treasury, Alexander Hamilton, the only other member of George Washington's administration whose influence rivaled Jefferson's.

With the hindsight of history, it has been customary to see Hamilton

Alexander Hamilton

and Jefferson as natural opponents: pragmatic banker confronting political ideologist, urban striver opposed to landed aristocrat, action man against ivory-tower thinker. They had both trained as lawyers and were imbued with the rational values of the Enlightenment. But while Jefferson was increasingly influenced by Jean-Jacques Rousseau's belief in the corrupting effects of wealth, Hamilton espoused the pragmatic Scots outlook of David Hume and Adam Smith that prosperity created civilization. It was no coincidence that Hume found nothing noble in a society of small farmers; he had seen Scotland move from rural poverty to commercial surplus and appreciated the difference. Like an Old Testament prophet, Jefferson warned that wealth poisoned society and that urban society was the most toxic of all; but Hamilton blithely accepted Hume's argument that prosperity allowed arts and refinement to flourish, and nowhere more obviously than in cities. "Thus," ran Hume's theme, "industry, knowledge and humanity are linked together by an indissoluble chain."

Eventually, the differences would force them apart, but at this stage Jefferson and Hamilton were colleagues and—almost—friends, who had seen at first hand the weaknesses of the Continental Congress and were determined to make the new Constitution's experiment with democracy work. It was an indication of their relationship that on June 20, 1790, Jefferson brokered a meeting between Madison and Hamilton that brought a famous bargain: the location of the nation's permanent capital on the Potomac River on Virginia territory, which Madison wanted, in exchange for the assumption of the states' debts, in particular their paper money, by the central government, the project on which depended all Hamilton's plans for the restructuring of the United States's finances. Soon afterward, Jefferson sent Hamilton a copy of his weights and measures report, and Hamilton replied in cooperative spirit. He had read the report with "much satisfaction," he said, and endorsed fully the concept of "a general standard among nations [which] seems full of convenience and order."

Later in the year, the first draft of Hamilton's own report on establishing a U.S. Mint appeared, and it contained a passage that would have warmed the secretary of state's heart. Referring to Jefferson's proposal to make the weight of the dollar equal to a new decimal ounce, Hamilton wrote: "There is an accuracy and spirit of system in this suggestion which constitutes a strong recommendation of it, if the advantage is not to be procured at the expense of some more considerable benefit. Fortunately there is such a coincidence of essential principles with this systematic idea, that it is not difficult to adhere to it."

However, the most remarkable instance of their harmony came in relation to the western lands, when Hamilton replied to a request from the House of Representatives for advice on ways to stimulate land sales, which so far had been disappointing. Unlike Jefferson, who saw the sales as an experiment in social engineering, Hamilton considered them simply as a source of funds to help pay the nation's debts. Most of the suggestions he offered in July 1790 were therefore designed to bring in more buyers—reducing the price per acre from one dollar to thirty cents, allowing two years' credit, and establishing additional land offices out on the frontier so that purchases could be made where the land was. Then he offered an unexpected helping hand to Jefferson's plan for decimalizing the measures.

The public lands should continue to be surveyed and laid out as a grid before they were sold, Hamilton reported, but in future the townships should be 10 miles square, containing lots measuring 100 acres and upward. He went no farther in specifying measures and dimensions but evidently did not consider it necessary. Jefferson's report had already laid out the values of the decimalized units of length, capacity, and weight that were being proposed for the United States. If Congress accepted those recommendations, the new units would necessarily be applied to the survey, and Hamilton's plan for decimalizing the survey would neatly accommodate the change.

What each man implicitly recognized was that the measures used in the survey of the public lands would also become the standard for the country as a whole. At the time the measures used were Gunter's, but any system could have been planted in the wilderness. There was, in Reverend Cutler's lambent phrase, no rubbish to be thrown out.

Consequently, Jefferson had grounds for confidence that his report would be well received. The Constitution placed responsibility for weights and measures with Congress. The members of that body now had before them a coherent plan that had the implicit backing of the president, of Hamilton and Jefferson, the strongest voices in the executive, and the indefatigable support of James Madison, one of the legislature's most influential members. It was what the hostile *Massachusetts Western Star* called an "overbearing and intriguing majority."

In the country at large most people were probably indifferent, but there was a small but vocal section of opinion in favor. As the *New York Daily Advertiser* declared, "This great desideratum in commerce and in social life . . . will probably be at length attained, and England, in conjunction with France, will perhaps have the honor of conferring this benefit on the rest of Europe." Pamphlets like John Beale Bordley's *On monies, coins, weights, and measures proposed for the United States of America* championed decimals and the ease, as Bordley put it, of "dividing by dots." Supportive letters appeared in New York and Philadelphia newspapers, and universities like Columbia and Pennsylvania lobbied for reform. Scanning the public and political landscape, the historian Julian Boyd, editor of Jefferson's *Writings*, estimated that "if ever a moment existed in which the public mind seemed ripe for a general reformation and in which political circumstances seemed auspicious, the summer of 1790 was assuredly that moment."

NINE

The Birth of the Metric System

THE FATE OF JEFFERSON'S PROJECT lay with the Congress, and at the first opportunity in December 1790, when Congress reconvened in Philadelphia, a committee was appointed to consider the secretary of state's proposals. For the converts, it had become a race to implement the new scientific system of measurement. Once a pattern of measurement and disposal of the western lands was established, they knew, it would become as resistant to change as the rest of the country. "Too long a postponement," Jefferson warned, ". . . would increase the difficulties of it's[*sic*] reception, with the increase of our population."

How short a time they had could be deduced from the pressure that Rufus Putnam was putting on the federal government to extend its power westward. In late 1790 Putnam wrote to Hamilton urging the United States to take over the Spanish-controlled Mississippi; otherwise the Northwestern Territory could not flourish. His frustration was born from the problems of the Ohio Company. That winter his family joined him, but they were still living inside Campus Martius. Marietta was growing fast and settlers had purchased lots far upstream on the Muskingum, pushing toward the flatter country that lay beyond the Alleghenies. But to become profitable, the company needed a surge of settlement, and the difficulty of shipping timber and produce down the Mississippi deterred would-be customers. With the

United States in control of the river, the last bar to settlement west of the mountains would be removed, and the population of the Ohio Valley would explode.

Despite living inside the fort, Putnam had no real fear of the territory's native occupants. Treaties had been signed at Fort Stanwix and Fort Harmar with the Six Nations and some of the Western Confederacy whose territory stretched from Michigan to the Ohio River. It was, however, recognized that care had to be taken. The surveyors who were the first into the territory had always suffered from having their horses stolen by Indians, and individuals traveling alone were at risk; but there was no immediate apprehension of greater danger. Then in the winter of 1790 a column of troops led by General Josiah Harmar was ambushed by Indians, and in January 1791 news reached the fort that fourteen people had been killed in one of the Muskingum settlements. Putnam communicated at once directly with President Washington.

"Our prospects are much changed," he wrote in his wayward hand. "In stead of peace and frienship with our Indian neighbours, a hored Savage war Stairs us in the face, the Indians appear ditermined on a general War."

Of the dozen or more nations in the Western Confederacy, those most affected by the Ohio settlements were the Delaware, Wyandot, and Miami. Although they hunted farther west in the winter months, in the spring and summer they cleared trees to plant corn, beans, and squash in the region around the foothills of the Alleghenies, into which Israel Ludlow and his surveyors were increasingly moving. Their hostility to the straight lines being scythed through the forest, and the growing number of square farms springing up in their wake, had grown to the point where it had at last burst into open warfare. In response, Washington ordered General Arthur St. Clair to take an army of 2,000 men to punish the Western Confederacy.

For the weights-and-measures reformers, the outbreak of hostilities came at a convenient moment. It held up settlement on the frontier, and in the Congress they needed all the time they could get. The Senate committee considering Jefferson's proposal for establishing uniform measures in the United States delivered its report in March 1791. Its membership included both James Monroe, a supporter, and the hostile Robert Morris, and in the end it was the latter who turned out to carry the most influence. The report noted that both France and Britain were considering how to obtain "a uniformity in the measures and weights of the commercial nations," and that as this "would be desirable," the United States should wait to see what the others would do.

The Senate had shown itself to be unwilling to take the lead, but the president swung his enormous prestige behind the object. On October 25, 1791, Washington returned to the need for reform, explicitly emphasizing the advantage of putting it on the scientific basis that Jefferson had argued for in his report. "A uniformity in the weights and measures of the country is among the important measures submitted to you by the Constitution," Washington declared in his annual address to Congress, "and, if it can be derived from a standard at once invariable and universal, must be no less honorable to the public councils than conducive to the public convenience." It was impossible to ignore such a command, and less than a week later the Senate appointed a three-man committee led by Ralph Izard to report on the subject.

Out in the Northwestern Territory, St. Clair was about to give the reformers another unexpected and bloodily won breathing space. With fatal overconfidence, he led his men up the Miami River from Fort Washington near Cincinnati, and on November 4 blundered into an ambush set by the Miami war chief Little Turtle. In the severest defeat the United States was ever to suffer in its Indian campaigns, 630 men were killed and another 270 wounded.

To buy time while another army was assembled under General Anthony Wayne, the secretary of war, Henry Knox, instructed Putnam to arrange a meeting with the Western Confederacy. "You will make it clearly understood that we want not a foot of their land," Knox ordered, none too honestly, "and that it is theirs and theirs only—That they have the right to sell and the right to refuse to sell."

Putnam did his best to follow these instructions at a council held in Vincennes in the summer of 1792, but he was never good at dissimulation. The Miami and most of their allies stayed away, and those who did attend clearly did not believe him. "I would have been glad if matters had remained as they were in the days of the French," said the chief of the Kaskaskia, Jean-Baptiste

Ducoigne. "Then all the country was clear and open. The French, English and Spaniards never took any lands from us. We expect the same of you." The chief of the Potawatomie recalled that he had refused a British attempt to buy his land, saying, "I foresaw that if I parted with my land I should reduce the Women and children to weeping." Each speaker made the same point, that the United States should advance no farther west. "It is best that the white people live in their own country and we in ours," said Ducoigne. "We desire of you to remain on the other side of the river Ohio."

Since there was no chance of that happening, the native inhabitants of the Northwestern Territory had to be made to change their minds. Only one body could do that. The 1787 land ordinance expressly reserved to the U.S. government the right to acquire land from the American Indians. But until Wayne's army was in a position to persuade them, further settlement west of the Ohio ceased. The pause had two effects: for several years settlers poured south rather than west; and the window of opportunity for reform of the United States's weights and measures remained open.

*T*HERE WAS NO MORRIS among the members of the Senate committee headed by Ralph Izard, but Monroe was still there. When the committee reported in April 1792, it was unanimous in recommending Jefferson's second scheme—for complete reform. It requested the president to have a second's rod constructed at public expense, adding that "the standard rod so to be provided shall be divided into five equal parts, one of which, to be called a foot, shall be the unit of measures of length for the United States. That the foot shall be divided into 10 inches, the inch into 10 lines, the line into 10 points; and that 10 feet shall make a decad, 10 decads a rood, 10 roods a furlong, and 10 furlongs a mile."

Everything Jefferson had wanted was there: "That measures of surface in the United States be made by squares of the measures of length; and that in the case of lands, the unit shall be a square, whereof every side shall be 100 feet, to be called a rood." The basic measure of capacity was a cubic decimal foot, to be called a bushel, and the basic unit of weight, the ounce, was to be derived as Jefferson had suggested from the weight of a cubic decimal inch of water.

The next step was for the Senate to adopt their recommendations. They showed no immediate urgency to act, but in the country the situation offered grounds for hope to the reformers.

The strongest opposition to any change came from the New York and Boston merchants most closely tied to trade with Britain; but even among that block of opinion there was a growing awareness of the need for more certainty in weights and measures than the traditional system could offer. To meet France's apparently insatiable hunger for wheat, for example, American exporters needed to buy from farmers outside the traditional growing areas in Pennsylvania and New York State. Each summer in the 1790s, the disputes that arose over payment for cargoes of grain brought home more forcibly than any political argument the disadvantage of having no agreed size of bushel. Simply in the interest of efficiency, a great tycoon like the diminutive John Swanwick, who had been Robert Morris's partner before creating his own business empire, was now prepared to accept that some certainty had to be introduced into the system.

What the reformers needed was some indication that at least one of the United States's major trading partners was about to make the same change.

$\mathcal{F}$OLLOWING TALLEYRAND'S RECOMMENDATION for reform in March 1790, the National Assembly had ordered the Académie to set up a commission on the subject. Among its distinguished members were such famous names as Borda; Pierre-Simon de Laplace, renowned as the founder of celestial mathematics and pioneer of probability theory; Antoine-Laurent Lavoisier, discoverer of the properties of oxygen; and, naturally, Condorcet. Its first report appeared in October, recommending that the system to replace the old aristocratic measures should be decimal. The arguments were those that Jefferson had marshaled: the simplicity of calculating in 10s, and the opportunity that this gave the average citizen of dealing on equal terms with those more educated than himself. Soon afterward, a new member, Gaspard Monge, an ardent Jacobin but also a mathematical prodigy who is now considered the father of differential geometry, was appointed to the Académie's commission. That winter it turned its attention to the choice of a scientific basis for the new system.

There were only three possibilities. The commission's report, published on March 19, 1791, listed them: "the length of the pendulum, a quadrant of the circle of the equator, finally a quadrant of the earth's meridian." As scientists, they clearly had to examine each possibility, but the report made apparent the clear advantages of the second's pendulum. The length was simple to calculate, and the results easily checked, so that the experiment could be run anywhere in the world without the need to go back to the place where the original trial was made. "In fact, the laws [of physics] concerning the length of the pendulum are sufficiently certain, sufficiently confirmed by experience to be used in experiments without fear of any but imperceptible errors." None of this was new, but the endorsement of such a distinguished panel was impressive confirmation of the method's reliability.

For the first eight pages of the French commission's report, the unity between the reformers in three different countries was maintained. Then on page 9, the commissioners sprang a bombshell. Having made the case for the second's pendulum, they suddenly commented: "However, we ought to observe that the unit thus derived contains something arbitrary. The second of time constitutes one eighty-six thousandth part of the day, and is consequently an arbitrary division of that natural entity. Thus, to fix the unit of length, requires not only a heterogeneous element—time—but one that is arbitrary." What the report implied was that a century of post-Newtonian research on the second's pendulum had overlooked a basic principle—time could not be used to determine length.

Having set aside the one internationally accepted foundation for a scientific system, the commission stated their preference for a unit taken from the earth itself, because this would be "analogous to all the real measures which in everyday life are also taken on the earth." They then swiftly dismissed the possibility of using the equator as a basis—too far away, too expensive, too complicated—leaving the meridian as the one remaining option. "In the end," they concluded almost frivolously, "one could say that everyone lives on a meridian, but only a part [of humanity] lives on the equator." At which point the report baldly declared, "A quadrant of the earth's meridian will therefore become the [basis] of the measure, and the ten millionth part of that length will be the unit used."

This, then, was to be the basis of the metric system, and it could hardly have been chosen in more capricious fashion. There was no scientific or rational basis to the argument against using the second's pendulum. Today the length of the meter is defined in terms of time—it is the distance traveled by light in a vacuum in a fraction of a second—and the fact that such a definition was, in the report's terminology, "heterogeneous" is irrelevant. All that the commission had to offer to justify its choice was the bald assertion that "it is a lot more natural to record the distance from one place to another in terms of a quadrant of one of the earth's great circles than to record it in terms of the length of a pendulum." Five of the finest scientific minds in the world, each trained in the merciless school of Cartesian dialectics, ought to have been able to come up with something more persuasive.

The speciousness of the commission's argument was immediately apparent to contemporaries and triggered speculation about the true reasons for its choice. The first clue came from a curious meeting between the scientists and Louis XVI in June 1791. By then, the *ancien régime* had collapsed, and power was gradually passing to more radical politicians and to the mobs demonstrating in the streets, forcing the king and his family to take refuge in the Tuileries palace in Paris. Given the surrounding danger, the reform of weights and measures could hardly have occupied many of Louis XVI's waking thoughts. One aspect of it, however, clearly snagged his curiosity.

Having chosen the meridian as its basis, the commission recognized that its distance would have to be measured accurately. Although it was not necessary to survey all ninety degrees from equator to pole to produce a reliable estimate, the longer the extent of it that could be measured, the more accurate the estimate; and to reduce errors to the minimum, both starting and finishing points needed to be at sea level. The line that suited all these requirements extended from Dunkirk through Perpignan to Barcelona. The length measured would cover slightly more than nine and a half degrees, enough to extrapolate with a good degree of certainty the full distance.

French geodesists had already triangulated most of it twice. For the new meter, the commission recommended that it be triangulated for a third time. Louis XVI wanted to know why.

On June 19, 1791, the Académie sent its most distinguished scientists to explain. That Louis decided to see them on that day throws light on a pleasing aspect of his otherwise rather dim personality. He was neither imaginative nor energetic, but he was interested in how things worked, and the depth of his interest can be inferred from the date. Alarmed by the growing power of the Revolution, and terrified that the Paris mob that prowled around the streets outside the palace would break in and lynch him and his family, the king had made plans to escape that very night.

Shortly after the scientists had left, the king wrote a long justification of the action he was about to take, then secretly changed into a servant's round hat and plain coat, while the queen, Marie Antoinette, disguised herself as a governess. At midnight on June 20 Louis slipped past the guards, followed some minutes later by the queen, and together they joined their children in a covered coach. The escape attempt almost succeeded. Despite the slowness of the coach, and an accident crossing a narrow bridge, they reached Varennes, barely three hours from the border, before a Revolutionary guard compared the servant's face with the image of the king on a banknote and recognized him. Louis and his family were brought back to Paris as prisoners, and he spent almost all the rest of his life under house arrest. Less than eighteen months later he was guillotined. Thus virtually his last autonomous act as a free man was to try to learn about the basis of the new decimalized system of weights and measures proposed for his kingdom.

When the scientists were assembled, Louis turned to Jacques-Dominique Cassini, the fourth of the family, who had been put in charge of measuring the meridian, and asked why he was going to repeat a measurement that his father and grandfather had already done. "Do you think you can do it better than they?" the king demanded.

Concealing the fact that he had already shuffled off responsibility for this physically demanding task onto two other scientists, Pierre Méchain and Jean-Baptiste Delambre, Cassini offered this answer: "Sire, I only flatter myself that I can do better because I have a great advantage over them. The instruments they had only measured angles to an accuracy of fifteen [degree] seconds. Monsieur le Chevalier de Borda has invented one which measures angles to an accuracy of one second. That is my entire justification."

The answer indicated that a shift of power had taken place, away from the idealistic, international dreams of Condorcet and toward the pragmatic ambitions of Jean-Charles Borda, the commission's chairman. This was confirmed by the physicist Jean-Baptiste Biot in his *Essai sur l'histoire générale des Sciences*, published in 1803, which revealed that Borda had persuaded the commission to select the meridian because it would enable them to achieve the goal of French science for more than a century, that of establishing the size of the earth. Since the meridian would have to be measured with the superb repeating theodolite that he had invented, the credit for completing the work begun by Picard, continued by generations of Cassinis, and involving expeditions to Peru and Lapland, would finally go to him.

The strange argument about the "heterogeneous element" of time was evidently supplied by Monge, who had his own radical agenda. In 1793 he became the driving force in the committee responsible for decimalizing the number of months in the year and days in the week, and for proposals to decimalize the hours in the day and, naturally, the number of seconds in the minute. Clearly, if the new measurement of distance were based on a pendulum ticking to the old second, it would have to be torn up when the new decimalized second was introduced. For Monge, the meridian meter was only the first step in a world made up entirely of 10s.

By itself, Borda's argument might not have been sufficient to persuade the commission to abandon the second's pendulum, but it was backed by another still more compelling motive, one familiar to generations of scientists and funding bodies since then: establishing the length of the meridian was a bigger, more expensive research project than establishing the length of the second's pendulum. On these grounds their choice was shrewd, for the National Assembly did indeed appropriate 300,000 *livres* for the project, a sum that kept many of the Académie's scientists in work when the institution was abolished in 1793 together with its annual grant of 93,000 *livres*.

There was a certain irony in this outcome, that it should have been Talleyrand, the most devious and dislikable of politicians, who gave expression to the high ideals of science, while the scientists abandoned their principles for materialist, personal, and political considerations. Nevertheless, on March 26, 1791, the National Assembly accepted the report of the Académie's commission in its entirety, and in a formal decree took the first step toward a break with the long history of organic measurement.

Once the meridian's distance had been established, one ten-millionth part of it would become the standard measure—eventually to be known as the meter—and from it would be derived a decimal system of lengths. At the same time, the decree ordered that two other experiments should be made: the first to establish the number of oscillations a pendulum of a meter's length would make in a day (this would serve as the means of verifying the results elsewhere in the world), and the second to establish the weight of a cubic measure of water—first a cubic *pied*, later the new decimal unit of a cubic decimeter, or tenth of a meter. This would provide the basis for the new decimal system of weights.

Thus what might have been an international enterprise became a French project. In London the commission's choice of the meridian tended to confirm ancient suspicions about France. Riggs Miller had failed to win reelection to the House of Commons, and in his absence the impetus for reform was fading. The government had already let it be known that it regarded the French commission's proposals for decimalization as "impracticable." Now it effectively turned its back on proposals so closely associated with the country that had always been Britain's rival.

*T*HE IMPACT OF FRANCE'S DECISION was felt most severely in the United States, where it cut the ground from beneath the feet of Thomas Jefferson. He had specifically tied his proposal to French science, deliberately abandoning measurement of the earth as a basis for his system, and going so far as to choose the latitude of Paris as the point where the definitive measurement of the second's pendulum should be made. "The element of measure adopted by the National Assembly," he now wrote in despair, "excludes, *ipso facto*, every nation on earth from a communion of measurement with them."

Despite the unanimous endorsement of his scheme by Senator Ralph Izard's committee, neither Jefferson nor Monroe could line up a Senate majority in favor of its recommendations. The president continued to press for reform, and enthusiasts for change could be found on each side of the divide that began to open up in 1791 as the legislature divided into two political groupings, Jefferson's Republicans and Hamilton's Federalists. But the consequences of Borda's last-minute switch hampered all their efforts. The existence of two proposals for reform encouraged more, and both Robert Livingston, later to be Jefferson's minister in France, and Oliver Wolcott, who became secretary of the Treasury under John Adams, put forward their own schemes.

It was soon apparent that the problem that concerned the Senate was not whether a change should be made but what form it should take. By the spring of 1793, it had considered the Izard report three times and recommended three different courses of action, but on each occasion it had accepted that the United States's weights and measures needed to have a scientific basis, and that the most convenient basis was the second's rod. After the last debate ended in deadlock, the exhausted senators voted to postpone further consideration of the subject and turned their minds to other matters.

Jefferson retired as secretary of state at the end of 1793, frustrated by the gathering power of the Federalists—and thereby earned Rufus Putnam's undying hostility for deserting the president. However, he never gave up his belief in decimals, nor did the campaign for reform end with Jefferson's departure. President Washington continued to support it, the Senate accepted the principle of change, and, as would soon become evident, opinion in the House of Representatives was also in favor.

Crucially, out on the frontier the window of opportunity remained open. Through 1792 and 1793, General Wayne's offensive against the Western Confederacy proceeded so cautiously that it was hardly apparent. During those years he constructed a string of forts to the west of the Indians' main force, so that they were gradually encircled. But so long as the Confederacy's warriors remained undefeated in the Ohio forests, no settlement of the Northwestern Territory could take place.

While the military situation still hung in the balance, a fresh impetus for change came from an unexpected direction. In France a new government took power in September 1792, when the former combination of king and National Assembly was replaced by a purely republican constitution. The temper of this new government—the Constituent Convention—was apparent in its decision to guillotine Louis XVI in January 1793 and, more important in the long run, to conscript 300,000 young men to fight in the army. Confronted by war on its frontiers and rebellion at home, the government set up the small, nine-member Committee of Public Safety with the power to coordinate every activity relevant to the country's security. Among these were the researches of the commission on weights and measures. The results of their work were now needed urgently: The old weights and measures were still in use, but because so many of the standards had been physically removed, uncertainty was giving rise to increasingly ugly disputes in the marketplace and at every rent-day.

Under pressure, the Académie's commission reported in the spring of 1793 with the recommendation that the new unit of length be named a *mètre*. Since the meridian survey was still incomplete, the commission used the results from an earlier triangulation of France to make a provisional estimate of the new unit's length. Their estimated figure for the meter's length was 443.44 lines, or 3 feet 11 inches and 44 lines. Monge's growing influence was apparent in the commission's other recommendations that time, geometry, and the calendar should all be decimalized as well.

On August 1, 1793, the Constituent Convention accepted the main points of the report—length to be based on the meter, area on the *are*, or 100 square meters, and capacity on a cubic decimeter (one-tenth of a meter). The basic unit of weight, named a *grave*—later to be renamed the kilogram—was to weigh as much as a cubic decimeter of water.

All this the scientists welcomed, but there was a sting in the tail. The new system, the Convention decreed, was to be in place twelve months later. It was an impossible timetable. The survey of the meridian from Dunkirk to Barcelona was less than a third complete—starting from the north, Delambre had barely reached the Loire, while Méchain, who had set out from Barcelona, was only just beyond the Pyrenees—and years would be needed to finish the work and to check the results. If the survey had to be abandoned, so too would be Borda's cherished project of establishing accurately the size of the earth.

Lavoisier protested furiously that, in adopting a slimmed-down speedy solution, they risked losing "a general system, which embraces geography, navigational skills, surveying, weights, currency and measurements of solids and liquids. In a word, it would mean losing, perhaps forever, the inestimable advantage of eliminating all calculation problems by the use of decimal divisions."

Antoine de Lavoisier

His response showed courage. The Convention had just abolished the Académie and was now beginning to direct its attention toward the Farmers-General, the tax organisation of which Lavoisier had been a member, accusing it of having made excessive profits at the expense of the taxpayer. In November, the all-powerful Committee of Public Safety ordered Lavoisier's arrest. This time Borda protested. "The presence of citizen Lavoisier is irreplaceable," he explained to the committee, "because of his unique talent for anything requiring the utmost precision. It is urgent that this citizen should be able to carry out the work he has always performed with as much zeal as energy."

Monge, however, stayed silent, and his reward came that winter when the Committee of Public Safety suddenly lost patience with the commission and declared that "responsibility should only be delegated to those who showed they were worthy of it by their republican virtue and their detestation of kings." Borda, Lavoisier, Laplace, Delambre, and others were dismissed, but Monge remained, the one real giant, however flawed. Deprived of its intellectual justification and of those who had brought it into being, the great experiment that was to unite all people for all time beneath the banner of science was on the verge of disappearing in confusion.

For scientists in particular, there must have been something particularly grotesque about the decision that winter to elevate reason to the status of a religion. The Académie had been abolished, Lavoisier thrown in prison, Condorcet threatened with arrest, and science in general was disparaged. In May 1794 a tribunal would send Lavoisier to the guillotine with the contemptuous remark "The republic has no need of intellectuals," prompting in turn Joseph-Louis Lagrange's memorably mordant eulogy, "It took them only an instant to cut off that head, but it is unlikely that a hundred years will be enough to produce another one like it."

That the idea of a decimalized system based on the meridian survived at all was mostly due to Maximilien Robespierre, the ideological perfectionist whose Jacobin party had sent Lavoisier to his death. Robespierre's radical rhetoric and command of bureaucratic procedure gave him control of the Committee of Public Safety, but his speeches suggested that personally he placed a higher value on his membership on the Committee of Public Instruction, whose mandate included not only education but anything that contributed to the general enlightenment of the citizen, including the new weights and measures. There could be nothing more purely rational than the metric system, based on science, universal in application, unencumbered by local variants, impeccably logical. "Surely it demonstrates," ran a Public Instruction report, "that in this field as in many others, the French republic is superior to all other nations." Adoption of the metric system was as patriotic a duty as wearing the three-colored cockade.

*I*N NOVEMBER 1793, Robespierre delivered a wide-ranging speech on foreign affairs, affirming France's solidarity with small, vulnerable republics, such as the Swiss cantons and the United States. There was little its conscript armies could do to aid the Americans, but French politicians were aware that the U.S. Congress was considering how to reform the country's weights and measures. Early in December, the Constituent Convention's president, Abbé Grégoire, suggested that it would be a friendly gesture to share France's intellectual achievement of the metric system with the Americans, and the idea was enthusiastically taken up by the Committee of Public Instruction. As its messenger, the committee chose Joseph Dombey.

Dombey was a fifty-two-year-old doctor, a botanist, and a man whose personality was as playful and charming and sensitive as a kitten's. There is little doubt that it was largely the combination of his character and his scientific background that led to his selection. "He was good-looking, with a charming smile," wrote his friend André Thouin, one of France's leading naturalists. "His eyes were large and dark, and the eyebrows thick and black. His skin was tanned, almost African in appearance. His character was gentle, open and sympathetic. In general he was high-spirited, although sometimes very low." He had integrity, courage, and a sense of adventure. He was the ideal choice in every way but one—his luck was phenomenally bad.

As a young man, Dombey had been a rising star in the founding science of botany, and in 1778 the head of the royal gardens selected him to join a Spanish botanical expedition to Peru and Chile. His five years there earned him a nomination to one of the forty-two places in the Académie. A herbarium of plants he sent back to Paris contained 1,500 new species and about sixty new genera. He brought back mineral specimens, including 38 pounds of platinum, discovered new species of hardwood in the headwaters of the Amazon, new shrubs in the Andes, and mercury and gold mines in Chile. He was attacked by Tupamaro Indians, collected Inca pottery, and described scientifically for the first time the magnificent 150-foot-tall Araucania pine.

Dombey should have been a hero, but pure misfortune destroyed all his achievements. One set of plants disappeared in a shipwreck, another was almost destroyed by Spanish officials, and, worst of all for a scientist, the Spanish government would not release his specimens until he had signed an undertaking not to write up his results until their own botanists had published theirs. In disgust, he gave up botany and returned to doctoring, but he chose to practice in Lyons, a city that Robespierre's Jacobins besieged and sacked in the summer of 1793. When the Jacobins entered the city, the doctor's patients were hauled out of the hospital for execution, the drains overflowed with blood running from the guillotines, and Dombey's mercurial spirits were so shattered that his friends thought he would go mad if he stayed in Revolutionary France. It was they who decided that he would make a suitable messenger to take the news of the metric system to the United States.

The idea quickly took root. Abbé Grégoire was told that Citizen Dombey, responsible for France's collection of South American plants, was anxious to create a similar collection from North America and was willing to travel there at his own expense. It might also have been pointed out that he was a particularly suitable choice, since he had brought back the platinum from which the definitive meter was to be fashioned. Grégoire was convinced.

On December 16, the Committee of Public Instruction issued instructions for a replica *mètre* and *grave* to be made out of copper for Citizen Dombey, and on the same date Robespierre and six other members of the Committee of Public Safety ordered that a passport be issued to Citizen Dombey so that he could travel to the United States in order to tell Congress of the advantages of the metric system. He would also send back to France useful plants and seeds and obtain information about the United States in response to a series of questions formulated by the Committee of Public Instruction and the Natural History Society.

Early in January 1794, the short, once-jaunty figure of Joseph Dombey, now growing bald and sallow-skinned, set foot on the gangplank of the American brig the *Soon*. In his baggage he carried the means of measuring the world.

TEN

Dombey's Luck

ON JANUARY 17 1794—the twenty-eighth day of Nivôse in Year II, by France's new decimalized calendar—Captain Nathaniel Williams Brown set sail from Le Havre with Joseph Dombey and his cargo, bound for Philadelphia. Swathed in linen and packed carefully in wooden boxes, the two copper objects Dombey carried with him looked like nothing useful, neither tool nor weapon nor kitchen utensil. One was a thin, flat bar, slightly more than a yard long, or as French citizens would have said, about half a *toise*. The other resembled the little cylinder an apothecary or a spice merchant might put on a set of scales to measure out his goods, except that it weighed slightly more than 2 pounds, a unit that no one on either side of the Atlantic had ever used. Locked into them was the most advanced science of the day.

Today, it is impossible not to recognize the significance of what was contained in the prototype meter and kilogram in Dombey's possession. They represented a watershed in human thought. That was what excited the French scientists above all else, the intellectual grandeur of the enterprise. Like the founding fathers, they were fully aware that they were making history. To devise a new system of weights and measures on a scientific basis involved a leap of imagination from the human-centered to the abstract. It was like inventing a new color. It was, Lavoisier had said, disregarding his own pioneering discovery of the properties of oxygen and of the composition of water, "one of the most beautiful and vast conceptions of the human mind."

Beside the precious objects and carefully folded into a leather pouch was a one-page document containing a decree issued in Paris on Frimaire 26 (December 16) and signed by, among others, Maximilien Robespierre, which explained the purpose of Dombey's voyage. It began, "The Committee of Public Safety [considers] that it would be useful for the United States Congress to learn of the work of the Committee of Public Instruction on weights and measures."

All that Captain Brown of New York would have known about his passenger was that his voyage had official authorization. Since the New York newspapers followed every move of the French Revolution in detail, seeing in it a rerun of their own nation's fight for freedom, someone as astute as a Yankee skipper could have guessed that Citizen Dombey, with his impressive passport, was probably an emissary of some kind from France's new rulers.

Dombey was a chunky, solid man, standing about 5 feet 4 inches, with a naturally lively face and, for all his hollow-cheeked fatigue, a charm that remained undiminished. Neither Brown nor Dombey had more than a smattering of the other's language, but events were to show that during the voyage the captain, like many before him, came to feel not just warmth but deep affection for the Frenchman. Whether he also grew to sense, with a sailor's instinct, the bad luck he carried with him was too small a detail to catch history's attention.

There is little doubt that in Philadelphia, too, Dombey would have been warmly welcomed. Despite the execution of the king, and the Terror that Robespierre had launched against the enemies of the Revolu-

Joseph Dombey

tion, France remained almost universally popular in the United States. It was still the ally whose ships, soldiers, and money had helped win the new nation its independence. Since the summer of 1793 and the outbreak of war between Britain and France, Britain had imposed a trade embargo against France, which had the effect of reviving memories of the recent Franco-American alliance. American grain ships heading for France were peremptorily halted on the high seas by the Royal Navy, and Congress resounded with furious speeches demanding retaliation. Even the secretary of the Treasury, Alexander Hamilton, the firmest supporter of commercial ties with Britain, turned hostile and gave the British ambassador a tongue-lashing about the "injuries which the commerce of this country had suffered." However patronizing Robespierre's view of the republic beyond the Atlantic might have been, he could hardly have chosen a more opportune moment to offer it the metric system as evidence of France's friendly feelings.

Although Jefferson was no longer in office, his French sympathies and eagerness to meet foreign botanists would have led him to make Dombey doubly welcome in Monticello. Dombey would have been supplied with introductions to James Madison, James Monroe, and Robert Livingston. He would doubtless have been bathed in that congressional sympathy for the French that found expression in April 1794 in fulsome motions from both houses "congratulating [France] upon the late brilliant successes of the arms of the Republic" against the British and their allies. President Washington, so concerned for the establishment of a uniform set of weights and measures to hold the young country together, would have approved the purpose of Dombey's visit and likely promoted his cause. The sight of those two copper objects, so easily copied and sent out to every state in the Union, together with the weighty scientific arguments supporting them, might well have clarified the minds of senators and representatives alike. The vibrant, determined personality of Dombey could have created an immediate empathy. And today the United States might not be the last country in the world to resist the metric system.

Dombey expected to reach Philadelphia at the end of February 1794, on the eve of his fifty-third birthday. He had escaped the carnage of the Terror, he was serving the cause of science, and he had been given a second chance at making his name as a botanist. When they were no more than a fortnight out from Philadelphia, a storm caught the *Soon* and drove it south toward the Caribbean, forcing Captain Brown to put in at the French colony of Guadeloupe for repairs. The island was sharply divided between royalists and revolutionaries. Fired by Dombey's presence, the leader of the revolutionaries distributed a poster attacking "that vile despot" General Victor Collot, the royalist governor of Guadeloupe. In response, the governor ordered Dombey's arrest as a troublemaker, at which a crowd of infuriated citizens threatened to attack a royalist village in retaliation.

As the mob advanced toward the Salt River, a sea inlet separating them from the village, Dombey's bedrock qualities of courage and fairness almost obscured the farcical nature of the misfortune that overtook him. Standing alone on the bank, blocking their crossing, he tried to reason with the mob and convince them that bloodshed was not the answer. When those in front would not listen, he shouted to those behind that he was prepared to go voluntarily to see the governor. Still the crowd would not turn back, nor would Dombey budge, until the physical pressure of bodies forced him off the bank. He fell several feet into the Salt River, and the surge of waves through the narrow inlet quickly swept him away from the bank. A boat was launched to pick him up. When he was eventually rescued he was unconscious.

For a few days more while he recovered from his near-drowning, Dombey continued to have the illusion of freedom, but in reality what would happen to him was inexorable. He had given his word to go voluntarily to Collot, and it was not in his character to break his promise. On March 26 he was taken around the coast to the island capital of Basse-Terre, where the governor took him into custody until his future could be settled.

With the *Soon*'s repairs complete, Captain Brown might have sailed for the United States. The Caribbean was a dangerous place. A British fleet under Admiral Sir John Jervis had just sailed into the West Indies, and British naval captains had a habit of stopping American ships to search for seamen who had deserted. They would certainly have removed a French national. There were pirates in those waters too and, more dangerous still, sea guerrillas, the privateers whom every maritime nation licensed to prey on the shipping of their enemies, and of neutrals suspected of helping them.

Dombey's presence on board would have been a liability, and it could only have been from friendship that Brown heeded Governor Collot's summons and on March 30 sailed around the coast to Basse-Terre to pick up his passenger. Recognizing the danger, Dombey disguised himself as a Spanish seaman—although it is hard to imagine what kind of pirate would have mistaken a careworn man with the pale skin and soft hands of a scientist for a sailor.

Brown was ready to sail the moment Dombey's chests were swung aboard, but he was forced to wait until a Swedish schooner under Collot's orders left the harbor first. Spectators on the steep slopes of the Soufrière volcano behind the harbor of Basse-Terre noticed that as the schooner reached open water, two privateers appeared over the horizon. Instead of turning away, the Swede altered course toward them and appeared to exchange some message before sheering off. Then the *Soon* emerged from the harbor, caught the wind in her sails, and steered north for the United States. The privateers did not even wait for her to lose sight of land. While the hillside audience watched, they closed in, forcing her to heave-to, and as she lay rocking in the swell a prize crew was sent aboard.

Some friends of the revolutionaries on Guadeloupe claimed to have seen the royalist governor Collot up there on the slopes as well. When the *Soon* was boarded, they alleged, he turned to one of his associates and remarked with satisfaction that Citizen Dombey was a plotter who had been taken care of. The British, he said, would certainly hang him.

That was not to be Dombey's fate. The privateers carried him and his belongings to the nearby British colony of Montserrat. He was a valuable prize, too valuable to hang. Among his papers were the diplomatic codes to be used by the French legation in the United States, for which the British would pay well, and Dombey was imprisoned in Plymouth, Montserrat's capital, until France was prepared to negotiate for his release. For the privateers, it was a good day's work. Captain Brown was soon returned to an American port, while the *Soon* together with her cargo, including the copper bar and weight, was destined to be sold as a prize.

Dombey did not survive long in his sunless stone cell. In early April he died and was buried on Montserrat. Today his grave, and much of Plymouth, is hidden under ash and rock thrown out when the volcano above the town began to erupt in the mid-1990s. Yet remarkably, the two measures that had cost him his life did reach the United States.

*S*OMETIME IN JULY 1794 the cargo of the *Soon* was auctioned off, probably in New York or Boston, and the prototype meter and kilogram, together with the message from the Committee of Public Safety, were bought by a French sympathizer, who sent them on to the French minister in Philadelphia, Joseph Fauchet. On August 2 Fauchet presented them to the new secretary of state, Edmund Randolph. Neither man was a scientist, and without Dombey to explain, they failed to appreciate the significance of the two standards. Indeed the prototypes of the meter and the kilogram were never shown to Congress at all.*

That winter, however, the Committee of Public Safety's message did reach the president. In January 1795 Washington communicated it to Congress with another appeal for action on the neglected question of weights and measures. Belatedly, the House of Representatives took up the challenge and appointed its own committee to consider both the president's appeal and Jefferson's scheme. In contrast to all previous reports, which had immediately recommended a new system, this committee restricted itself to declaring some basic and by now familiar principles: that change was needed, and that a new system should be derived from the second's pendulum and the cube of rainwater. It did not commit itself to a decimal system but simply asked that $1,000 be allocated to conduct experiments to determine which new system would be most convenient.

Unlike the Senate at that period, the House of Representatives allowed its debates to be published, and the opinions expressed in its discussion of

---

*Somehow they survived two centuries of bureaucratic neglect, and by the sort of good fortune that escaped their bearer, came to rest in a display case in the National Institute of Standards and Technology (NIST) in Washington. NIST is the government department that spearheads official efforts today to bring the metric system to the United States. In that sense, Dombey's copper bar and weight have still not arrived.

Dombey's meter (left) and kilogram (right).

the report give a good idea of those in the public at large. Some were for change and some against. Others abstained, saying that only scientists could understand the problem. When Jonathan Havens of New York earnestly tried to explain the science behind the second's pendulum, his fellow New Yorker William Cooper jumped to his feet and mockingly declared that it put him in mind of the lines from Oliver Goldsmith's *The Deserted Village*:

> *While words of learned length and thundering sound*
> *Amaz'd the gazing rustics rang'd around;*
> *And still they gaz'd, and still the wonder grew,*
> *That one small head could carry all he knew.*

A sense of gravity was restored when the tycoon John Swanwick of Pennsylvania pulled himself up to his scant five feet and said that he had known of too many "disputes which have arisen for want of some certain standard to regulate weights and measures, and frequently the payment of a whole cargo [has been] disputed on account of a difference in the size of bushels." Thomas Jefferson had always accepted that the business community was most hostile to change, and this call for reform by a founder of the Bank of New York, and an unquestioned leader of the city's financial interests, represented a significant shift in attitudes. Swanwick was supported by Jonathan Dayton, who wielded a double influence as land speculator and Speaker of the House. The subject of weights and measures was very important, Dayton insisted, "and to no country more than the United States, as every state has its own different weights and measures which causes the greatest uncertainty in all commercial transactions." The commercial argument was one every congressman understood. The jeers and dissent were silenced, and the House adopted the report unanimously.

Quite suddenly, it appeared that the French example had, however belatedly, given the political process the kickstart it required. A bill was submitted and passed by the House. Under the title "An Act directing certain experiments to be made to ascertain uniform standards of weights and measures for the United States," it was sent to the Senate. There it sailed through two readings and the committee stage, and on May 31, 1795, the last day of the session, the Senate informed the House that it would consider the bill at the next session.

At which point it vanished. The next session came and went, and the bill on weights and measures was never seen again.

THE EXPLANATION FOR ITS DISAPPEARANCE lay with events taking place far away from Philadelphia, in the forests beyond the Ohio River. After two years of cautious fort construction far to the west, where the Alleghenies start to flatten into the Indiana plains, the U.S. army under General Anthony Wayne had succeeded in encircling the forces of the Western Confederacy. In the summer of 1794 this army advanced northward along the Maumee River toward the shores of Lake Erie,* and in a clearing made by a tornado on the west bank of the river, they encountered the Confederacy's

*Near the modern city of Toledo, Ohio.

warriors. From defensive positions behind the tangle of storm-felled trees, Wayne's troops fired almost unscathed into the exposed positions of the Indians until return fire slackened, and then with a final bayonet charge won an emphatic victory.

The Battle of Fallen Timbers led the leaders of the Wyandot, the Delaware, the Miami, and other Indian nations to sign a peace treaty the following year at Fort Greenville. There the Confederacy was forced to acknowledge U.S. sovereignty over its land and to allow settlement up to a line from the mouth of the Kentucky River to Lake Erie.

Two other treaties were signed in that year of 1795: the Jay Treaty with the British, who finally agreed to evacuate Detroit and other territory on the northern frontier; and the Treaty of San Lorenzo with the Spanish, which granted Americans freedom of navigation on the Mississippi and the use of New Orleans as a port. Once the menace of Indian attack, of British harassment, of Spanish blockade was removed, the logjam that had held up settlement at the Ohio River suddenly gave way. Behind the outriders of surveyors and squatters came an irresistible army of settlers who were to push the Wyandot, the Miami, and their allies out of Indiana, out of Illinois, into one treaty after another, farther and farther west, until their identity almost disappeared.

For ten years, ever since Jefferson had first thought of the strategy, the reform of weights and measures had been linked to the sale of public land. Hamilton acknowledged implicitly what Jefferson had declared explicitly, that so long as the country remained virtually unsettled, it might be measured in any fashion and newcomers would accommodate themselves to the new units. Once it was populated, the possibility of change disappeared. That was the significance of the Battle of Fallen Timbers: With the defeat of the Western Confederacy, the window of opportunity for introducing a new system had closed.

On May 18, 1796, displaying an alacrity conspicuously absent where weights and measures were concerned, Congress passed "an Act for the sale of land of the United States in the territory northwest of the River Ohio, and above the mouth of the Kentucky River." The period of experimentation was over. The land was to be surveyed by the government before sale, and this time there were to be no variations in the procedure—no squares alternating between townships and sections, no land company sales. The public survey's definitive pattern of 6-mile-square townships, divided into 1-mile-square sections, was established for good. Lines would run north-south, and east-west, and one section in each township would be reserved for education.

"The lines will be measured," the act carefully specified, "with chains containing two perches of 16½ feet each, divided into 25 equal links, adjusted to a standard kept for that purpose." The fact that a perch was also a pole and a rod might be confusing, but the authoritative gloss given by C. Albert White in *A History of the Rectangular Survey System* makes the act's import clear. "This specifically calls for a Gunter's chain, and leaves no doubt that accurate measurements are to be made." On a wider scale of significance, this was also the first U.S. unit of measurement to be designated in law. Congress might not have been able to make up its mind about any other weight or measure, but it was determined that the dimensions of Gunter's chain should apply to the entire nation.

On the frontier and elsewhere, decimals were dead.

*I*N JUNE 1846, Charles Dickens traveled from London to Switzerland to begin work on a new novel that had a plot but no title. His journey took him through Mâcon, the town in eastern France where Joseph Dombey had been born. The family still lived and worked there, and the name, possibly glimpsed on their grocery store or in an advertisement for it, must have lodged in Dickens's memory, because a fortnight after arriving in Switzerland he wrote delightedly to a friend, "BEGAN DOMBEY!" The unbending character of Mr. Dombey in *Dombey and Son*, coldly ambitious for himself and his family, could not have been farther from that of the warm, mercurial Joseph, but so far as the wider world is concerned, the fictional Dombey is the only one who has ever existed.

ELEVEN

The End of Putnam

THE *Pittsburgh Gazette*, a reliable guide to the economic health of the Ohio Valley, reported in October 1795 that "the emigration to this country this fall surpasses that of any other season—and we are informed that the banks of the Monongahela are lined with people intending for the settlement on the Ohio, and Kentucky. As an instance of the increasing prosperity of this part of the state, land that two or three years ago was sold for ten shillings [$2.50] per acre, will now bring upwards of three pounds [$15]."

To investors in the Ohio Company who had bought at about twelve cents an acre in 1785, the surge of settlers pushing up land values at last brought healthy profits. For Rufus Putnam in particular, it represented a triumph of planning and persistence, and in November 1796 he was rewarded by being appointed surveyor-general of the United States, with responsibility for surveying the rest of the Northwestern Territory.

In the aftermath of the Greenville Treaty, the great fort at Marietta was torn down and its red timbers were used to construct more elegant, peaceful dwellings, the finest of them housing Putnam and his family. Looking around their home today, a visitor would find no incongruity between the belongings they had brought from Massachusetts—the fine bow-fronted walnut desk, the splendid four-poster bed, the deep-toned cello—and the handsome building with its dark, gleaming floors and its smooth, dazzlingly white plaster walls. Refinement had reached west of the Ohio.

Then in his late fifties, and seemingly as energetic as ever after the treaty, Putnam personally undertook the survey of the reserves that the Moravian missionaries had persuaded the government to set aside for them and their converts among the American Indians. He remained brusque and straightforward in his work, conducting the immense task of surveying the Northwestern Territory from a plain, two-roomed land office built at the back of his house. He contracted with Israel Ludlow to run the line marking the boundary specified by the Greenville Treaty between the Western Confederacy and the American settlers. It zigzagged toward the southwest from the mouth of the Cuyahoga on Lake Erie to the mouth of the Kentucky River on the Ohio. Ludlow, as might have been expected, managed to bend the line by more than 500 yards. Reporting this to Putnam, he expressed the hope that the Indians would not notice, and since both agreed that it was a risk worth taking, the mistake remained uncorrected.

In surveying terminology, the crook in Ludlow's line was known as a jog, as though his elbow had been nudged while drawing it on the paper. The plats produced by the teams of surveyors working for Putnam were filled with jogs. He had divided the huge area northwest of the Ohio River—effectively, central and western Ohio today—into districts and assigned them to different survey teams, but the east–west parallels in one district rarely matched up with those of its neighbors. The north-south meridians added to Putnam's difficulties. Although the 1796 act had specified that the meridians were to be run true north, the business of getting a fix on the polestar, and correcting for its deviation and for local magnetic variation, took too much time. None of Putnam's teams bothered with it, and he did not insist. Consequently, the enormous Military Reserve, for example, which the United States had set aside for its veterans in Ohio, was tilted four degrees off true north. Nor did anyone have time, at three dollars a mile, to solve the problem of converging meridians, and sporadic corrections meant that sections that were supposed to contain 640 acres varied in size from under 600 acres to more than 700.

Although the federal lands were surveyed in squares, the territory of

Ohio was distributed in no fewer than nineteen different grants to states, veterans, religious and other organizations. Putnam surveyed the external boundaries, but then each owner could use its own measuring methods to divide up its grant. Consequently, Ohio served as a surveying laboratory. The land that Connecticut had reserved in exchange for its western claims was surveyed in squares with sides 5 miles long. On behalf of its veterans, Virginia had reserved four million acres, which were surveyed by metes and bounds. Symmes's purchase was surveyed using magnetic rather than due north. Significantly, the only real failure was the Virginia Military Reserve.

Metes and bounds allowed the claimant to present his bounty warrant and designate the shape and location of the parcel of land he wanted. So long as the land had not already been claimed, it was then surveyed and registered as his. In the Virginia Military Reserve it quickly became apparent that all the best land had gone to the first-comers, and as the late arrivals tried to fit themselves in, the shapes to be surveyed became increasingly complex. Soon the reserve resembled a jigsaw puzzle. One parcel ended up as a polygon with 118 sides, while another, supposed to contain only 458 acres, was discovered to measure 1,662 acres. Both then and now, the huge majority of land disputes in Ohio have concerned these irregularly shaped parcels, and all over the United States, wherever the metes-and-bounds system was practiced, it produced similar results.

By 1798 nearly all the land inside the treaty line had been surveyed. Some of the jogs were ludicrously large, but a pattern of (roughly) 6-mile-square townships had been laid across the land, and the bristly old soldier Rufus Putnam had established a system of issuing contracts for specific surveys that all his successors were to follow.

Other surveyors took the same pragmatic attitude, that precision was less important than getting the work done. In 1796 Moses Cleaveland, surveyor and major stockholder in the Connecticut Land Company, was running straight lines through the area that stretched north of Thomas Hutchins's Seven Ranges to Lake Erie. This was the 3.5-million-acre Western Reserve that Connecticut had cleverly required the federal government to award it in return for giving up its charter claims in the West. The land company had bought it for $1.2 million, a sum that in New England fashion was earmarked for education, and proved to be the foundation for Connecticut's nineteenth-century public education system—but the company first needed a return on its investment.

It was all good land, as the description of the Moravian missionary John Heckewelder made clear. "Although the country in general containeth both arable Land & good Pasturage," he wrote, "yet there are particular Spots far preferable to others." The best of all in his opinion was the place where the Cuyahoga River flowed into Lake Erie. It had a safe harbor, and the fishing there was good, Heckewelder reported, because it was "a place to which the White Fish of the Lake resort in the Spring in order to Spawn." The river itself "has a clear & lively current but all Waters & Springs emptying in the same prove by their clearness & current that it must be a healthy Country in general." Cleaveland was in search of a suitable location for the capital of the Western Reserve and, attracted by Heckewelder's magical description, took a surveying party and began to mark out a city among the trees.

For anyone unaware of Gunter's chain, the dimensions of what would eventually be Cleveland, Ohio, are hard to understand. The largest road, Superior Street, measured 2,640 feet long and 132 feet broad, while minor streets were 99 feet in width. But for chainmen working in dense woodland, it made sense to keep things simple, hence a main street 40 chains long by 2 chains broad, and lesser roads 1½ chains across. The main square was 10 acres in area, which happened to be 10 chains by 10 chains, and most of the city lots were 2 acres in area, or, in surveyors' terms, 2 chains broad by 10 chains long.

Wherever speed was necessary, surveyors would "gunterize" the measurements in this fashion—the "National Road" constructed on Congress's orders across central Ohio and Indiana in 1796 to link settlers with their markets in the East was also 99 feet broad—because it was quicker to keep things simple for the chainmen. For them, as for Putnam, the priority was finishing the job rather than being fancy.

In Ohio, no one thought the worse of the U.S. surveyor-general for his strange spelling, or for putting his son and a son-in-law on the federal payroll, or for his standards of surveying. He was regarded with affection and respected as someone who got things done. With the beginning of the new century, however, it became clear that in the public lands survey this rough-

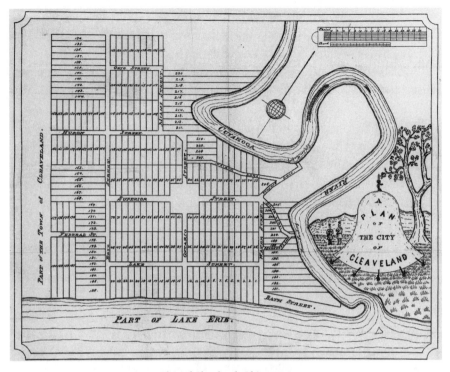

Plan of Cleveland, Ohio, 1796.

and-ready approach would no longer be tolerated. After the election of Thomas Jefferson as president in 1800 and his appointment of the ferociously clever, Swiss-born Albert Gallatin as secretary of the Treasury, the messages from the nation's new capital, Washington, to Marietta became noticeably peremptory in tone.

Putnam had never forgiven Jefferson for his leadership of the Republicans, which he regarded as disloyalty to President Washington. The increasingly hostile attitude of his administration confirmed Putnam's prejudice. Then in 1803 Gallatin fired him—"because," Putnam commented bitterly, "I did not die nor resigne"—and without acknowledging his achievements brusquely instructed him to turn over his papers to the new surveyor-general, Jared Mansfield, a former professor of mathematics at West Point military academy.

Putnam responded with dignity, handing over the records but writing to Gallatin in a tone that offered a hint of the lion on his coat of arms: "Perhaps you may imagine this conduct looks like passive obedience and non-resistance, or that I am courting favor. Mistake me not. I have done no more than what I conceive to be the duty of every public officer. . . . I am too independent to be influenced by the prejudices of the times."

It was Putnam's last public growl as surveyor-general. Surrounded by grandchildren and soothed perhaps by the sound of Persis's cello, he passed the remainder of his days in Marietta. By the time he died in 1824, four years after his wife, the land survey had been pushed westward across Indiana and Illinois and stretched from Lake Michigan south to the Gulf of Mexico.

The changes introduced during those years by Putnam's two successors, Jared Mansfield and Edward Tiffin, showed why he had had to be replaced. It was not just the prejudice of the times that led to his sacking; it was a fear that the entire federal land survey stood in danger of collapse.

*T*HE DANGER STEMMED from the uncontrollable surge in land speculation. More than in any other economy of the time, American land was the prime producer of wealth, partly from its crops and livestock, but mostly from the increase in its value. The math was spelled out crudely by a spokesman for the North American Land Company, which had invested in unoccupied land west of the Appalachians. The population had doubled in the last twenty-five years, the spokesman said in 1790, and would again in the next twenty-five. "Supposing half of each state to be unoccupied," he explained, "it follows that in twenty-five years there would be an increase of inhabitants sufficient to settle this vacant half. The average price of the lands in the settled half of the United States cannot be at less than eight dollars an acre—sixteen times the price at which the lands of the company are [bought] for its shareholders."

The figures were tempting, and everyone with spare cash invested in land. Washington did, Supreme Court justices did, congressmen, senators, and governors did. "All I am now worth was gained by speculations in land," the new secretary of state, Timothy Pickering, told his sister in 1796. "In 1785 I purchased about twelve thousand acres in Pennsylvania which cost me about one shilling [about fifteen cents] in lawful money an acre. . . . The lowest value of the worst tract is now not below two dollars an acre."

One of the heaviest plungers was Judge James Wilson of the Supreme Court, who at the urging of his friend Silas Dean, an associate of Robert Morris, took every chance to acquire land, however dubious the circumstances. "If we review the rise and progress of private fortunes in America," Dean advised him, "we shall find that a very small proportion of them has arisen or been acquired by commerce, compared with those made by prudent purchases and management of lands."

The trick was not to wait for demand to push up the price, but to buy, as the Ohio Company did, with devalued paper currency. Although the Continental Congress had been responsible for the majority of notes and warrants printed during the war, the states had continued to issue their own loan and treasury certificates after peace came. In the 1780s, most could be bought on the street for twenty cents on the dollar, and by one estimate almost three-quarters of North Carolina's paper money was held by a group of speculators small enough to meet in a single room. When the states began to put their own public lands on sale, the holders of such notes could use them to pay for land at full value. It was with these state sales rather than the disposal of federal land that the corrosive effects of speculation really began to bite.

The double goal of raising money and attracting settlers into previously unoccupied regions was what led states to sell their public lands. Once the surveyors had run lines establishing their borders, each individual state started to put territory up for sale. Between 1783 and 1800 the states sold around fifty million acres, far more than the federal government, and the same names appear repeatedly as buyers: Robert Morris, and his unrelated namesake Gouverneur Morris, Alexander Macomb, William Duer, and William Bingham of Pennsylvania.

Free enterprise was born out of land dealing, and long before the first business corporation existed, land companies issued shares and created many of the financial and legal structures that the nineteenth-century stock-dealing, capitalist economy used to finance the railroads and industrialization of the United States. As early as 1765, Patrick Henry created the first pure trust, the North American Land Company, a legal ploy he devised to protect the assets of none other than Robert Morris. Such devices became more common after the Revolution, and financial institutions in Boston, Philadelphia, and New York issued prospectuses and evolved increasingly sophisticated financial structures to tempt people to invest in land. Many of the largest transactions were handled by banks in Britain and the Netherlands, which had access to sums up to one million dollars, far beyond the reach of any ordinary settler.

There was nothing illegal about buying up military warrants in whatever quantity. And despite the public outcry, a court found no overt evidence of bribery after Alexander Macomb persuaded the New York legislature to sell him more than three million acres for as little as eight cents an acre. Nevertheless, the line between speculation and corruption was becoming blurred. It began to disappear altogether at the point where northern financial sophistication mixed with southern metes and bounds.

A METES-AND-BOUNDS SURVEY did not just produce shapes that only the best surveyors could measure; it created a maze of bureaucratic form-filling that invited fraud and wholesale corruption. In theory, the purchaser handed over to the state treasurer the money or military warrant for the amount of land he wanted. The receipt was taken to the land registry, which issued another warrant, which had to be taken to the county surveyor, who laid off the land that was wanted and gave the purchaser a certificate describing the property. The certificate was returned to the land registry so that the property could be patented, and only then was a document finally issued proving that the purchaser was indeed the owner of the land.

In practice, the procedure was complicated further by the inaccurate maps drawn up by poorly trained surveyors, and by the mistakes made—freely or for bribes—by inadequately paid registrars and land officers so that legal claims were forgotten or predated. Sometimes disputes were settled by force on the land itself, sometimes in court, and occasionally by free admission, as in 1816 when Kentucky's auditor contritely revealed that hundreds of legally purchased farms had never been registered, and that as a result the

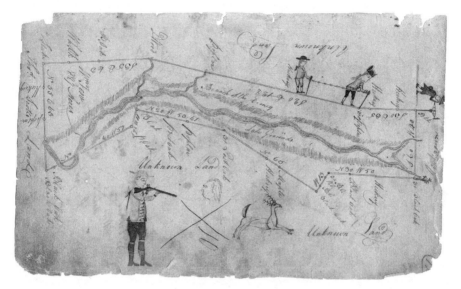

Metes-and-bounds survey in Georgia, 1785.

state had sold the owners' land all over again. Absentee buyers found it advisable to employ local land-jobbers to protect their investments by running fresh surveys, registering their claims, and suing in the county courts. The qualities they looked for were those of Uria Brown, a well-known Virginia jobber who in 1816 promised his clients that he would "appear solid & firm, & persist in Establishing the rights of Lands as the Tuffest skin shall hold out the Longest . . . & surveys on surveys is there, [k]nee Deep and deeper."

In effect the metes-and-bounds system was skewed in favor of those with deep enough pockets to hire lawyers and land-jobbers, and to keep sweet an army of state officials. These were the very people who could afford to invest in depreciated warrants or paper money. Robert Morris had shown how it was done in 1783 with his speculative purchase of more than one million acres in Virginia, paid for in military warrants. When more land was released by the state in 1792 in exchange for heavily discounted Virginia treasury notes, the speculators descended like vultures and bought up the notes, backed by credit from brokerage houses in Boston, Philadelphia, and New York. A total of 2.5

million acres ended up in the hands of just fourteen individuals, most of whom were absentees. George Washington himself pointed to the consequence in western Virginia, where "the greater part if not all the good Lands, on the main river, are in the hands of persons who do not incline to reside thereon themselves, and possibly hold them too high for others."

The result was that huge swaths of the southern states—up to three-quarters of the southern Appalachians, according to modern research—ended in the hands of absentee landowners. Much of what remained was owned by a tiny aristocratic minority. Just 4 percent of the land, the most mountainous and least productive part of it, remained to be divided up among 80 percent of the population, the majority of whom were left as landless tenants, sharecroppers, or slaves.

So much empty soil belonging to absentees was a recipe for resentment. Landless residents burned the blazed trees that served as metes-and-bounds markers, or moved monuments, and occasionally entirely new plats were forged. To substantiate their claims, they relied on local knowledge, and on homegrown specialists known as "red brush surveyors" who knew every wrinkle of the land and could recognize boundaries by memory. But unlike the northern states, southern legislatures, dominated by the interests of landowners, allowed no rights of preemption to squatters or "trespassers having no color of title" until the 1820s.

The often violent struggles that broke out between the absentees and the jobbers trying to hold on to their vacant land, and the squatters and red brush surveyors trying to move in, had an unintended consequence. Doubts over who actually owned properties slowed the market in land. In contrast to the North, where families would settle, improve, sell, and move on in a restless fury that astonished European visitors, southerners, once they had bought their land, tended to stick there, at least partly because legal doubts restricted the land market to local buyers and sellers. In 1816 Uria Brown warned that speculators "would purchase no Lands in [west] Virginia at any price: for the Titles of Land there was worse than the Titles in Kentucky." And to show how bad that was, he prophesied, "The titles in Kentucky w[ill] be Disputed for a Century to Come yet, when it's an old Settled Country." It proved to be a sound prophecy. Generations of Kentucky lawyers have earned fame and fortune from the endless battles over land originally measured out by metes and bounds.

The consequences of those distant surveys and their associated complications continue to mark the rural economies of southern Appalachia. Modern studies of eighty counties in the area show that with uncanny persistence almost three-quarters of owners continue to be absentee, and over half of the 20 million acres are owned by one percent of the tax-paying population. Power and money still rest with these giant holdings, now often in the hands of mineral, timber, and agricultural concerns, that occupy the most productive ground. The contrast with the subsistence farmers barely surviving on small holdings up in the high woodland is as stark as it was two centuries ago.

A more fruitful legacy lies in the relationship of southerners to their land, which often seems closer than in the North. A farm whose boundaries are natural—the trees, streams, and ridges chosen by metes-and-bounds surveyors—should inspire warmer feelings than one defined by sharp lines and rectangles, but if it cannot be sold, it may also come to feel more like a prison. Sometimes sublime, sometimes tortured, especially where the property has stayed in the family for generations, such feelings infuse the work of writers like William Faulkner and Eudora Welty. "Don't you see?" exclaims Ike McCaslin in a fine gothic frenzy in Faulkner's *Go Down Moses*. "This whole land, the whole South, is cursed, and all of us who derive from it, whom it ever suckled, white and black both, lie under the curse?" The impact of metes-and-bounds surveys and flawed land titles on southern literature would be worth studying.

In Georgia, speculation crossed the blurred line into outright fraud. Alone among the original states, Georgia had refused to cede to the United States its charter claims to land in the West, which included much of present-day Alabama and Mississippi. The Appalachian mountain range that separated other states from their western territory only skirted the northwest corner of Georgia, and there was no physical barrier to prevent settlers moving west until they reached the Mississippi. Consequently, the state, the youngest and poorest of the thirteen, was eager to sell the land before it was taken up by squatters. Ever since South Carolina's plantation owners had moved across the border in colonial days and helped break down James

Oglethorpe's surveyed plan of settlement, Georgia's land sales had been anarchic. They were about to become fraudulent on a massive scale.

In 1789 three different companies were formed by syndicates from Tennessee, South Carolina, and Virginia—this last headed by Patrick Henry—and persuaded the Georgia legislature to let them buy twenty million acres centered on the Yazoo River valley for $207,000. Most of the territory belonged to the Cherokee, Choctaw, and Chickasaw peoples, and according to federal law could only be bought by the U.S. government. The audacious deal failed because the companies wanted to pay in depreciated certificates although the state demanded cash. Nevertheless, the seed was sown.

In 1795 Spain relinquished its claims to Georgia's western territory, and four new companies were promptly formed to acquire land now undeniably American. Within months both houses of the Georgia legislature passed legislation selling them forty million acres, much of modern Alabama and Mississippi, for just $500,000, although only half that sum was actually paid. An official inquiry by James Madison later discovered that all but one of the legislators had an interest in the sale, either in shares or in bribes.

Public outrage voted in an anti-sales party, which overturned the agreement thirteen months after it was made, but the damage had been done. Yazoo land had already been sold on for more than $1.5 million; among the secondary purchasers were Judge James Wilson of the Supreme Court, who bought 750,000 acres; William Blount, governor of the Tennessee Territory; members of the U.S. Congress; and, almost inevitably, Robert Morris. Claiming that Georgia's agreement to sell the Yazoo lands was a valid contract, they took their case to court, and eventually to the Supreme Court. In 1809 Chief Justice John Marshall, a friend of Robert Morris and an advocate of company rights, held that the sale, despite its fraudulent circumstances, was indeed a contract. The case, Fletcher vs. Peck, not only won the Yazooists compensation but underpinned the entire land market that was to fuel the United States's westward expansion.

The sheer scale of the Yazoo fraud has prompted one modern analyst to compare Georgia at that period to modern Russia, where powerful cartels aim not only to break the law but to use it to legitimate their criminal activities. Their corrupting influence was not confined to the state. A Territory governor, federal and state legislators, and justices up to the U.S. Supreme Court had all indirectly benefited from the fraud, and it was the federal government's alarm at what was happening that led to Madison's inquiry. One consequence of his findings was that in 1802 the United States brought pressure to bear on Georgia and at last persuaded the state to cede its western claims to the federal government. Another was that Albert Gallatin and Thomas Jefferson decided the United States needed to have a surveyor-general whose standards of accuracy could not be questioned. That was why rough and ready Rufus Putnam had had to go, and why Jared Mansfield had been appointed surveyor-general of the United States.

*W*HAT JEFFERSON AND GALLATIN FEARED was that fraud and confusion would spill out of South Carolina and Georgia and into the federal public lands. The 1790 law creating the Southwestern Territory— which consisted of Tennessee and, after 1802, of Alabama and Mississippi— required the land to be surveyed, sold, and governed on the same basis as the Northwestern Territory. The practice was different. In the North, something like a grid was being imposed on public land; in the South, metes and bounds had become the norm, giving rise to a climate of corruption, litigation, and a land market that rewarded the speculator and the banker. For a president whose political philosophy centered on widespread ownership of land, it was intolerable.

The surveyor of the South, Isaac Briggs, a talented engineer who would soon be involved in the successful survey of the Erie Canal, evidently froze at the tangle of problems before him. The heat, the fever, the belts of heavily wooded, swampy terrain, all made surveying so difficult that even at a special rate of four dollars a day he could not find enough men prepared to take on the work. The surveyors he did hire attempted to lay out townships on the northern model but found they had to deal not only with squatters but with a slew of existing properties bought through the Yazoo land companies and other speculative investors, or by private arrangement with Spain, or, especially in regions like Natchez, Mississippi, where many Americans had acquired Cherokee land, by deals with the native inhabitants. Each had to be separately surveyed and entered on the plat—"a severe duty which the surveyors complain of," explained Briggs's successor, Thomas Freeman, in 1807. There were so many that whatever the law might demand, their plats had the disordered appearance of a metes-and-bounds survey.

Determined to prevent southern land patterns from taking over, Gallatin wrote impatiently to Briggs in 1805: "It is true that you will not be able to complete your work in that scientifick manner which was desireable, [but] it is of primary importance that the land should be surveyed and divided, as well as it can be done." It was advice that would have drawn a grim smile from Rufus Putnam.

The date of Gallatin's letter was significant. A few months earlier, the last detail had been completed on the Louisiana Purchase, a deal that added 900,000 square miles of new territory to the 400,000 already in the public domain. It was no longer a matter of having a proper survey of Alabama and Mississippi, but of the gigantic expanse of land the United States now owned west of the Mississippi River.

The Louisiana Purchase occurred because the priorities of both sides made a perfect fit. Napoleon had acquired the land from Spain just three years earlier but needed money to finance the next phase of his European war. His goal was matched by the United States's readiness to pay good money to control the port of New Orleans at the mouth of the Mississippi. No doubt some deal would have been reached in any circumstances, but the sheer scope of what was eventually agreed on must have been partly due to the familiarity of the French negotiator with the workings of the American land market.

In 1794 Talleyrand had traveled to New York, fleeing from Robespierre's campaign of terror in France. As an apostle of French metric reform, he ought to have visited Thomas Jefferson, but he had a stronger tie with Gouverneur Morris. Before the French Revolution, Gouverneur had gone to France to negotiate tobacco deals for his business colleague and namesake Robert, and for a time had shared Talleyrand's mistress, Adèle de Flahaut. Fashionable circles used to speculate on the sexual orientation that led her to choose Talleyrand, who had a clubfoot, and Gouverneur, who stumped around on a wooden leg; but none of the three cared. "My friend's counte-nance," Gouverneur wrote of Adèle, "glows with Satisfaction in looking at the Bishop and myself as we sit together agreeing in Sentiment & supporting the Opinions of each other."

Naturally, in the United States Talleyrand attached himself to Gou-verneur Morris and thus entered Robert Morris's circle. He soon found him-self dispatched to Maine to spy out new lands for purchase there, and as he surveyed the grandeur of the landscape his thoughts became unmistakably American. "There were forests as old as the world itself," he wrote, "green and luxuriant grass decking the banks of rivers: large natural meadows, strange and delicate flowers. . . . in the face of these immense solitudes we gave vent to our imagination. Our minds built cities, villages and hamlets." Failing to make a living from real estate, Talleyrand eventually returned to France and by 1803 had risen to become Napoleon's foreign minister.

Having worked for Robert Morris and dreamed the speculators' dreams, Talleyrand knew the background from which the American negotiators came. Robert Livingston, the American minister in Paris, owned the massive Clermont estate on the upper Hudson and knew every large speculator in New York, and James Monroe, who arrived in April 1803 as Jefferson's per-sonal representative, was a piedmont Virginian steeped in a tradition of land acquisition. Consequently, when Talleyrand suddenly offered American ne-gotiators not just New Orleans but all of France's territory beyond the Mississippi, the bizarre nature of what was happening—the sale of an em-pire—was not an issue; it was simply a land deal. The sum asked, fifteen mil-lion dollars, exceeded by five million the maximum that Monroe and Livingston were authorized to spend, but that was less important than the price per acre. Although neither side knew the exact size of French Louisiana—"I can give you no guidance, but you have made an excellent deal for yourselves," Talleyrand said airily—the two Americans could work out that with public domain land selling at two dollars an acre there would be enough profit for them to close the deal without referring it back to Jeffer-son. (It turned out that they paid less than five cents an acre.) The financing for the transaction was provided by Alexander Baring of Barings Bank in

The Louisiana Purchase.

London, who, as the son-in-law of William Bingham, owner of some four million acres from Maine to South Carolina, was as familiar as anyone with wheeling and dealing in American real estate.

Once it belonged to the United States, the Louisiana Purchase needed to be surveyed and sold off. The land beside the Mississippi was divided into Orleans Territory, comprising the region round New Orleans, and Louisiana Territory, which covered modern-day Arkansas and Missouri. But when the public land survey moved into the area, every problem that Isaac Briggs had encountered on the east bank of the Mississippi was to be found on the west. The United States had undertaken to honor property registered with the French and Spanish authorities, but the promise constantly threatened to break the land survey. In Orleans Territory, the long lots of French farmers, measuring 5 *arpents* by 40 *arpents* (about 320 yards by 2,600 yards) and running back from rivers and creeks, were so numerous they forced surveyors to abandon 6-mile squares altogether wherever there were waterways. Silas

Bent, the surveyor-general of Louisiana Territory, found more sinister problems.

Wherever there was good land, squatters claimed to have French or Spanish land grants, and they not only blazed trees but also forged the records. The registry office files "have undergone a revolution," Bent reported to Gallatin in 1806. "There has been Leaves cut out of the Books and others pasted in with Large Plats of Surveys on them. . . . the dates have been evidently altered in a large proportion of the certificates. Plats have been altered from smaller to Larger. Names erased and others incerted and striking difference in collour of the ink etc."

There was a large and growing threat that all the corrupt practices and inefficient surveys that dogged land distribution in the southern Appalachians would become endemic west of the Mississippi. The entire administration of the public lands survey had to be overhauled. Surveys had to be closely controlled, their accuracy improved, and a system of checking imposed. Above all, the standards expected of the surveyors had to become professional. That was the challenge facing the new surveyor-general of the United States, Jared Mansfield.

TWELVE

The Immaculate Grid

THE REGULARITY THAT JARED MANSFIELD imposed on both himself and the United States was ruthless. From the moment he left Yale in 1777, his life as teacher, soldier, and instructor at West Point was one of such blameless rectitude, it is almost a relief to discover that he was in fact expelled from Yale for "discreditable escapades." Yet even here his crime—cheating on an exam—might be seen as an attempt to achieve perfection. Certainly, the undeviating straightness of the rest of his career must have served as some form of compensation for the one stain on his character.

Mansfield was a mathematician of the highest class, and his *Essays, Mathematical and Physical*, published in 1802, was the first contribution to that field by a native-born American. Astronomy and the calculations needed to establish earth positions from celestial observation received particular emphasis in his writing. It was noteworthy that, following his appointment as surveyor-general, Mansfield placed an order with leading instrument makers in London for "A three-foot Reflecting Telescope, mounted in the best manner, with [William] Wollaston's *Catalogue of the Stars*, [Nevil] Maskelyne's *Observations and Tables*, A thirty inch Portable Transit Instrument, answering also the purpose of an Equal Altitude Instrument and Therdolete [*sic*], [and] An Astronomical Pendulum Clock." This was not the action of someone who would tolerate curving straight lines and squares that looked like diamonds.

The best way to appreciate what Mansfield did is to drive west from Dayton, Ohio, on Interstate 70 and just across the Indiana state border to swing north on State Route 277. Beyond the wide expanse of the interstate, a beautiful rolling landscape can be glimpsed between predatory trucks and glaring billboards, but once on the narrow strip of 277, you become part of that country. The road acts as boundary to fields of wheat and soya, lawns with picket fences back onto it, and it is shaded by stands of oak and maple and walnut. The geological tsunami of the Alleghenies has virtually subsided here, leaving only ripples of rock beneath the surface. Sometimes blacktop, sometimes gray gravel, 277 runs straight as a surveyor's rule over the ripples, giving a ride as exhilarating as a powerboat at sea. In the troughs, all you can see are the nearest red-painted barns and green John Deere harvesters, but from the peaks the distant spires of churches and silver grain towers appear above the surging land.

However, Route 277 has another claim to distinction. Exactly one mile to the east, Indiana's border with Ohio runs parallel to it, a north-south line that Jared Mansfield designated as the First Principal Meridian. West of that line he was to establish a survey whose squares were so immaculate that their pattern would be compared to graph paper, checkerboards, and plaid. In other words, Ohio, with its different and indifferent surveys, was the proving ground for the system, but 277 marks the first line inside Mansfield's monumental gridiron.

Its regularity grew from the initial point formed at the meeting of the principal meridian—a carefully surveyed north-south line—and an east-west baseline crossing it exactly at right angles. The squares were numbered outward from that zero point: Running east-west they were called ranges, and north-south they were townships; thus the first square west of a meridian and north of a baseline would be titled Range 1 West, Township 1 North. Mansfield himself personally surveyed the Second Principal Meridian, whose initial point can still be found a few miles south of Paoli, Indiana, and others were run as new areas of land were pried from the Native Americans and put on the market. Instead of Rufus Putnam's independent survey districts, the different areas could be connected by extending a baseline or a meridian. The most spectacular example was the Fifth Principal Meridian, whose initial point was in Arkansas, near the present town of Blakton, 26 miles west of the Mississippi, but which was extended so far north that it eventually controlled other land surveys in Arkansas, Missouri, Iowa, South Dakota, North Dakota, and most of Minnesota, ending with Township 164 North, on the Canadian border.

Mansfield's method also provided a solution to the problem of converging meridians. The curvature of the earth brings lines of longitude gradually together as they run toward the pole, so that in most of the United States the northern end of a township is 30 to 40 feet narrower than the southern. In Alaska the flattening of the earth means that the lines close by more than 100 feet. (Some of Israel Ludlow's township meridians in Ohio converged by more than 300 feet, but that was not entirely the earth's fault.) As townships were stacked on top of each other, they grew narrower, and after four or five, the most northerly might be over 60 yards narrower than it was supposed to be. Consequently, Mansfield's successor, Edward Tiffin, decreed that a fresh start should be made at that point, with new meridians exactly 6 miles apart marked off on the baseline. The jog created by these "correction lines," where the old north-south line abruptly stopped and a new one began 50 or 60 yards farther east or west, became a feature of the grid, and because back roads tend to follow surveyors' lines, they present an interesting driving hazard today. After miles of straight gravel or blacktop, the sudden appearance of a correction line catches most drivers by surprise, and frantic tire marks show where vehicles have been thrown into hasty ninety-degree turns, followed by a second skid after a short stretch running west or east when the road heads north again onto the new meridian.

Despite these improvements, Mansfield's surveyors were not immune from error. Massive jogs occurred in eastern Illinois, where the parallels running out from the Second Principal Meridian spectacularly failed to meet up with those of the Third Principal Meridian, and a "shatter zone" of steeply angled lines had to be introduced to join one lot to the other. But such mistakes were inevitable given the pressure the surveyors were under to make land available for sale as quickly as possible.

Rapidly though the squares were laid out, it was not quick enough to satisfy the press of settlers. The westward drive was fueled partly by the rising population—the 1800 census showed that in ten years the number of Americans had grown by 35 percent to 5,306,000—but especially by the mouthwatering prospect of the land beyond the Alleghenies, beginning roughly at Chillicothe, Ohio, whose ancient mounds proved that it had been productive ground centuries before the first European landed. For the land-hungry, timber-yearning, field-dreaming squatters in Kentucky and Virginia, those gentle slopes, rich meadows, and rolling oakwoods beyond the hills were the stuff of dreams. Once clear of the mess of claims and counterclaims around the city of Dayton, this country stretched out empty and inviting as far as the sterile, unforested prairies.

As early as 1798, the federal government in Philadelphia had sent Rufus Putnam an anxious request for information about a group of 300 Kentucky families who had settled in unsurveyed country just beyond the treaty line. When he went to investigate, he found that many had come intending to buy the lands "as they should be offered for sale by the United States," but that rising prices had pushed them on. Even in Dayton the price of two-dollar town lots with uncertain title were, according to the speculator John Symmes in 1796, "selling in Cincinnati at ten dollars per lot." Cheaper land could only be found farther west. The race that developed between the surveyors and squatters marked the entire history of the land survey, and it was rare for a surveying team to measure productive country that had no settlers at all.

In their desire for land, pioneers like John Pulliam, who had already squatted on farms in Virginia and Kentucky, leapfrogged far ahead of the tide. In the breakout year of 1796 he crossed the Ohio, and with his family moved on through what would become Indiana and Illinois as far as the Mississippi River. Much of this was compacted prairie soil, where tall bluestem grass and wild rye grew taller than a horse's head, and their roots knotted into an unyielding mass that could not be broken up until John Deere's heavy, self-scouring steel plows were introduced in the 1840s.

The Pulliam family was Scotch-Irish, and as stubborn and restless as its eighteenth-century forebears who had swarmed unchecked through the Virginia piedmont in the 1730s. With their few goods piled in a cart, accompanied by long-legged hogs and some scrawny cattle, they settled for a year or two on land beside any creek where there was enough water to produce a stand of timber. Even shallow-rooted elms and sycamores broke up the ground well enough for it to be plowed, and provided timber for making cabins. Along with maples, whose sweet sap in springtime could be boiled into sugar, there were oaks and beeches that produced acorns and mast for the hogs, and hickories like the pecan whose nuts could be pounded to flour or eaten whole, and fruit trees and wild vines carrying ripe cherries and sweet grapes. This produce ensured that those first settlers were timber dwellers rather than grassland farmers.

The story of the Pulliam family was collected by nineteenth-century antiquarians because John's descendants were the first to move into Sangamon County in Illinois, but as John Mack Faragher commented in *Sugar Creek*, his 1986 history of that community, it is likely that John Pulliam "spent his whole life farming without ever owning land." The reason was simple. The 1796 land act specified that the land was to be sold by the section, that is, 640 acres, and at a minimum price of $2 per acre, which, with costs for the first year's subsistence and improvements like fencing and building, left little change from $2,000. To someone like John Pulliam, whose last days were passed as a ferryman on the Kaskaskia River, this was an impossible sum.

Successive pieces of legislation rapidly reduced the smallest amount of land that could be bought. In 1800 it was a half section (320 acres), and the down payment was only a quarter of the total price, the rest being paid over the next three years. Congress authorized the sale of land by the quarter section (160 acres) in 1804, and in 1820 by the half-quarter section (80 acres), with the price reduced to only $1.25 an acre. The changes had their effect, and Robert Pulliam, John's son, took the opportunity to buy the land that he farmed around Sugar Creek in Sangamon County. A shooting accident cost Robert his right foot, and his temper grew evil, but that did not prevent him acquiring more acres than his father ever squatted on.

The beauty of the land survey as refined by Jared Mansfield was that it made buying simple, whether by squatter, settler, or speculator. The system

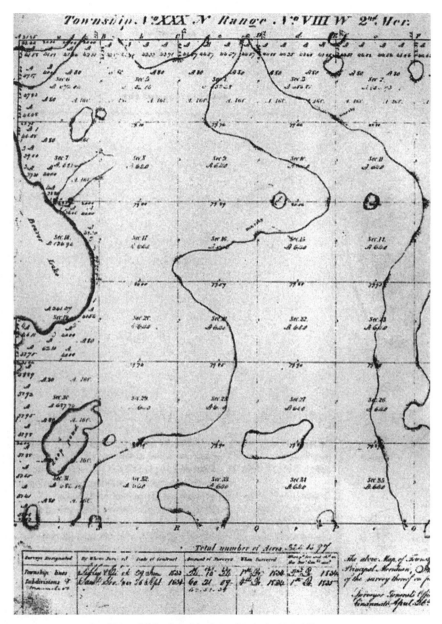

Jared Mansfield's plat of the Second Principal Meridian, 1836.

gave every parcel of virgin ground a unique identity, beginning with the township. Its name might be Township 2 North, Range 4 West, Second Principal Meridian. Within the township, the thirty-six sections were numbered in an idiosyncratic fashion established by the 1796 act, beginning with Section 1 in the northeast corner, and continuing first westward then eastward, back and forth in "boustrophedonic" fashion, that is, like an ox pulling a plow, until Section 36 was reached in the southeast corner.

Each square-mile section consequently had its own identity, and as the size of the minimum parcel of land shrank, its identification simply grew more specific. In 1832 the smallest area for which a would-be farmer could bid at a government land auction was reduced to a quarter of a quarter-section, or 40 acres, a parcel that has entered American rural mythology. The surveyors hated it, complaining bitterly of the paperwork involved in "the new and minute subdivisions of fractional sections," but in the jargon of modern geographers it would become "the modular unit of settlement." It was the minimum area that was needed to support the average family. Railroads sold land by the 40-acre lot. After the Civil War, freed slaves were reckoned to be self-sufficient with "40 acres and a mule," and in the twentieth century real-estate developers preferred to deal in 40-acre parcels. Even rotary sprinklers were designed to irrigate 40 acres of grass. And long before the U.S. Postal Service ever dreamed of zip codes, every one of these quarter-quarter sections had a specific name, such as "¼ South-West, ¼ Section Northwest, Section 8, Township 22 North, Range 4 West, Fifth Principal Meridian."

What the system depended upon was a clear title to the property. This was indeed the very purpose of the grid. As a result, there was always a strand of opinion that regarded squatters as a menace. In 1807 a draconian act authorized their punishment with fines and imprisonment, but it was rarely enforced, and after years of turning a blind eye to the practice, the Preemption Act of 1841 gave squatters the legal right to do what up to then had been a custom backed by local squatter power: buy the land they had improved by plowing or building a cabin, at the going rate of $1.25 per acre. But long before then, most squatters had recognized the power of a surveyor's plat.

The desire to possess land drew people westward, but it was the survey that made possession legal. Three years after Robert Pulliam began farming in Sugar Creek, the deputy surveyor for that district, Angus Langham, arrived with his chainmen and axmen, who were also referred to as moundmen in the woodless prairies because they constructed mounds rather than blazing trees at section corners. These were a different generation from the surveyors who had worked for Rufus Putnam, and their growing professionalism could be seen in the instruction booklet issued to each of them.

First the township or exterior lines were to be established. From the initial point, the surveyors were instructed to move westward along the baseline parallel across the entire district, marking it with posts at every half-mile and mile for the quarter-section and section lines, as well as with 6-mile markers for the township corners. From each of the 6-mile markers a meridian was then run north, marked in the same way. Finally, the team returned to the initial point and worked its way north along the principal meridian. At each 6-mile point, an east-west parallel was run to cut across the north-south lines. Branded onto every corner mark was the number of the section, the township, and range.

Along the way surveyors had to make a map on a scale of 4 inches to the mile, and keep notes in their field-book, recording the distance covered and the principal natural features found. And in case they did not know, the manual offered this example of how to record a meridian running north from a baseline:

North Along the east boundary of Section 36, Township 21 north of the baseline, Range 6 east of the 4th principal meridian.

Chains

14.70   *A brook, 25 links wide, with a rapid current, runs south westerly about 10 chains, then turns to the N.W.*

27.60   *Left the creek bottom and entered hills.*

29.40   *A white oak, 15 inches diameter.*

33.70   *A hickory, 24 inches diameter.*

40.00   *Set a quarter Section corner post on the top of a ridge, bearing north easterly and south westerly; from which post a white oak bears S[outh] 28 [degrees] W[est] 197 links, and a poplar*

> *bears N[orth] 56 [degrees] W[est] 14 links distant. The soil is*
> *good and fit for cultivation; timber walnut, cherry and white*
> *oak; undergrowth pawpaw and spice.*

49.07   *A white oak*

64.08   *A walnut*

80.00   *Set a post, corner to Sections 25 and 36, Township 21 north,*
> *Range 6 East of the 4th principal meridian; from which a*
> *hickory bears South 57 degrees West 127 links; and a white oak*
> *bears North 23 degrees West, 72 links distant. Land too hilly*
> *for cultivation, although the soil is rich; timber, hickory, white*
> *oak and walnut; undergrowth pawpaw and spice.*

That represented one side of a square-mile section, and the same procedure had to be repeated for all 36 square miles of the township. Then, beginning at the southeast corner of each township, they worked west and north
filling in the section lines. Shrewd speculators learned to avoid land in the
northwest corner of a township, because any errors in measurement showed
up there, and that was where most arguments over boundaries occurred.

When the surveyor's plat was delivered to the district land office, it
would show that every square of prairie or corner of forest had been given an
identity; where Robert Pulliam's farm stood it was Section 21, Township 14
North, Range 5 West, Third Principal Meridian, together with the letters *AP*,
meaning "applied for," which showed that the land had been preempted.
Eventually, the claim had to be paid for, and once that was done it was entered on the definitive survey map and patented, at which point the prairie
or forest became private property, whose ownership would be protected by
the full force of the law. This was something the wildest squatter eventually
learned. As a means of defense, a surveyor's plat came to carry more fire-
power than a Kentucky long rifle packed full of lead shot and black powder.

Once that was understood, the influence of the grid stretched out to
places long before the surveyors arrived. Anyone who intended to settle and
buy land would try to mark out his claim in rectangles so that it aligned with
the grid. In 1832 one nameless pioneer who had moved far ahead of the survey into the Iowa prairies explained how he did it: "The absence of section

lines rendered it necessary to take the sun at noon [to find north-south] and at evening [to find east-west] as a guide by which to run these claim lines. So many steps each way [800 double paces by 1,600] counted three hundred and twenty acres, more or less the legal area of a claim. It may readily be supposed that these lines were far from correct, but they answered all the necessary claim purposes for it was understood among the settlers that when land came to be surveyed and entered, all inequalities would be put right [by adding or subtracting land]."

What this makes clear is how easy it was to measure out land using the grid and the old 4-based organic measures. Acres had evolved from the basic productive needs of human society, and in the primitive conditions of the frontier they again came into their own. When Congress made it possible to sell public land by the quarter-quarter section, the squatter had only to step out 250 double paces, or 440 yards, toward the sun at noon, stick in a marker, then march 250 double paces toward the point where the sun set—"only alternate steps are counted," W. F. Horton reminded readers of his *Landbuyer's, Settler's, and Explorer's Guide*, published in 1902—and put in another marker. That was it: 440 yards by 440 yards made a 40-acre lot. Even Jefferson would have approved; no system of measurement could have been more transparent, more democratic, more suited to "the calculation of everyone who possesses the first elements of arithmetic." It was so straightforward that the citizen squatter could operate it as easily as the government surveyor.

*A*S TERRITORY SOUTH OF TENNESSEE was squeezed out of the Cherokee, Choctaw, and Chickasaw, Mansfield's system was applied there as well. Two principal meridians were established, the Washington Meridian in southeast Mississippi, and the St. Stephen's Meridian in southern Alabama, both taking as their baseline the thirty-first parallel, the boundary between the United States's and Spain's territory. Yet even under Mansfield's rigid rules, northern regularity still battled to overcome southern customs. Squatters measured out their land not by squares but by metes and bounds, and the survey's rectangles had to be fitted round these irregular parcels. West of the Mississippi River, repeated claims to having owned land under Spanish or French rule forced surveyors to make stringent checks for falsified records and newly cut blazes on trees. But the very nature of southern agriculture ran counter to the rigid pattern of squares.

In the first two decades of the nineteenth century, cotton prices doubled, and plantation owners in Virginia and the Carolinas began to sell their exhausted land and move to Alabama and Mississippi accompanied by their slaves, livestock, furniture, and machinery. It was not like a pioneer family moving west. This was an undertaking on the scale of relocating a factory today and was often financed by investors in New York and Philadelphia. Every care had to be taken to acquire the best site, and the soil most suited to cotton growing was the rich valley land around Natchez and along the lowlands east of the Mississippi River. Most of this was owned by the Cherokee, although some was squatted, or the claims poorly documented. To acquire legal title to it amid the confusion of existing private claims required inside knowledge. Wealthy purchasers soon found they could buy this information from the public land office in Huntsville, Alabama. In 1818, a year after General John Coffee was appointed surveyor-general of Alabama, the *Huntsville Republican* informed its readers that the land office would "give any information to people wishing to purchase an advantage," in return for "a liberal per centum" of the price. Whether the client wished to locate a suitable tract or to engross several different parcels, if necessary removing squatters, Coffee's employees, the newspaper alleged, were prepared to "do business on commission, and receive in pay either a part of the land purchased; or money."

The society that evolved from this pattern of land distribution seemed far less exotic to foreign visitors than the one in the North. The South's large estates and social privilege produced a hierarchy that nineteenth-century Europeans found familiar, except for its reliance on slavery. Many were shocked by the discovery that slavery was still legal in the United States. What they failed to appreciate was how far the concept of property had evolved in the new nation. Where slaves were regarded as property—"For actual property has been lawfully vested in that form," Jefferson bleakly wrote in 1824, "and who can lawfully take it from the possessors?"—that concept overrode their humanity and the founding principle that all men were created equal with the inalienable right of liberty.

The 1787 ordinance's ban on slavery—at the insistence of Rufus Putnam—meant that the new states in the North had escaped that evil, and in the northerners' energetic, egalitarian materialism, European visitors found the utterly foreign experience many of them were looking for. It was based on owning the land that Mansfield had measured out for the settlers. "The possession of land is the aim of all action, generally speaking and the cure for all social evils among men in the United States," wrote the visiting English writer Harriet Martineau in *Society in America* in 1837. "If a man is disappointed in politics or love, he goes and buys land. If he disgraces himself, he betakes himself to a lot in the west. If the demand for any article of manufacture slackens, the operatives drop into the unsettled lands. If a citizen's neighbors rise above him in the towns, he betakes himself where he can be monarch of all he surveys."

It was so obvious, so widespread, that few Americans recognized that it was remarkable, but European visitors immediately found in the pattern of land ownership something quite distinct from the hierarchical, landlord-dominated, agricultural societies they were accustomed to. One of the earliest accounts, from John Melish, a political radical who first traveled through the West in 1806, shows clearly that he saw landownership as the key to American independence of spirit. "Every industrious citizen of the United States has the power to become a freeholder, on paying the small sum of eighty dollars, being the first installment on the purchase of a quarter of a section of land," he wrote in *Travels in the United States*; "and though he should not have a shilling in the world, he can easily clear as much from the land, as will pay the remaining installments before they come due."

This was optimistic. Land office records were filled with entries of farmers who had failed to keep up their payments. But Melish did not let such detail obscure the picture he wanted to paint of a society unlike anything known to his readership in Britain. "The land being purely his own, there is no setting limits to his prosperity. No proud tyrant can lord it over him—he has no rent to pay—no game laws—nor timber laws nor fishing laws to dread. He has no taxes to pay except his equal share for the support of the civil government of the country, which is but a trifle. . . . Such are the blessings enjoyed by the American farmer. . . . May the Almighty Father of the human race pour down his choicest blessing on the heads of those who planned, and carried into effect such a benevolent system."

Not surprisingly, Jefferson read Melish's book "with extreme satisfaction." It takes a small effort to appreciate how odd the American system would have appeared to most of Melish's readership, in whose experience the privilege of owning land freehold was largely confined to the aristocrat, the squire, and the gentleman. "I *own* here a far better estate than I *rented* in England," wrote Morris Birkbeck in 1818, "and am already more attached to the soil. Here, every citizen, whether by birthright or adoption is part of the government, identified with it, not virtually but in fact. . . . I love this government." Birkbeck wrote to pull in immigrants to the Illinois colony he had founded, but the emotional connection he made between land ownership and democracy was unmistakably genuine. This was the bright side of the coin called property.

When Captain Frederick Marryat, a bluff naval officer and prolific author of best-selling adventure stories, visited Galena, Wisconsin, in 1837, he could hardly believe that such privileges of ownership were supported by the law even against the government, even when they were claimed by a squatter. As an example, he wrote of the federal depository that had been built at Galena to hold the lead that miners in the territory paid as tax. "As soon as the government had finished it," he recorded, "a man stepped forward and proved his right of pre-emption on the land upon which the building was erected, and it was decided against the government, although the land was actually government land." Marryat's tone was amused rather than admiring, much as when he wrote of the reaction of backwoods squatters to attempts to evict them, "the consequences were very commonly that the new proprietor was found some fine morning with a rifle-bullet through his head." It was all part of the wild and woolly West, he implied, and not something that any well-run society would want to copy.

The French politician and writer Alexis de Tocqueville was so struck by the system of landownership in the United States that he made the mistake of asserting in *Democracy in America* (1835–40) that "in America there are, properly speaking, no farming tenants; every man owns the ground he tills." In fact, much land was leased or rented, but on a small scale compared to the

Alexis de Toqueville

pattern of tenant farmers in Britain, and landlord and peasant in France and Germany.

Unlike John Melish, the wonderfully named English writer Fanny Trollope, mother of Anthony, was a Tory to the tip of her parasol, but from the opposite end of the political spectrum her conclusions matched his. They were prompted by a visit to a pioneer farm in the forest near Cincinnati in 1828. The family lived in a log cabin they had built beside a river, with a peach and apple orchard at the back, and cows, horses, and pigs nearby. They grew potatoes on land they themselves had cleared, wove and knitted their own cotton and woolen clothes, made soap and candles, went to the market to sell butter and chickens, and buy tea, coffee, and whiskey, and were, she admitted, "indeed independent." But they had achieved this rural idyll, she decided, at the cost of cutting themselves off from social life as she understood it: "They pay neither taxes nor tythes, are never expected to pull off a hat or to make a courtesy, and will live and die without hearing or uttering the . . . words, 'God save the king.'"

The social consequences of such independence were quite clear to her, as she wrote in *Domestic Manners of the Americans* (1832): "Any man's son may become the equal of any other man's son, and the consciousness of this is certainly a spur to exertion; on the other hand, it is also a spur to that coarse familiarity, untempered by any shadow of respect, which is assumed by the grossest and lowest in their intercourse with the highest and most refined."

*F*OR GOOD OR ILL, a new kind of society was evolving from the way in which the public land was being measured out. It could be seen in Congress, where a perceptible change in policies and attitudes occurred as representatives of the new western states, like the future president William Henry Harrison, a delegate of the Ohio Territory, took their places.

In his influential 1893 essay *The Significance of the Frontier in American History,* Frederick Jackson Turner identified this change as the point when the United States took on the individualistic, egalitarian qualities he associated with the frontier spirit: "The frontier promoted the formation of a composite nationality for the American people. . . . In the crucible of the frontier the immigrants were Americanized, liberated, and fused into a mixed race." As a result, frontier people had a sense of themselves as American and an instinct for democracy that they had to teach the Atlantic states, which were still ridden with sectional self-interest. "The rise of democracy as an effective force in the nation," Turner asserted, "came in with Western preponderance under [Andrew] Jackson and William Henry Harrison, and it meant the triumph of the frontier."

Attractive though it is, Turner's argument has been repeatedly demolished by historians and geographers pointing out its inherent inaccuracies. Westward expansion was always piecemeal and irregular rather than consisting of a coherent moving frontier. Far from being individualistic, it was usually communal, and often built around urban rather than rural settlements. Above all, it was fueled more by speculation than by the desire for liberty. Yet for all the academic scorn, the thesis refuses to die, simply because a distinctive, utterly American spirit did indeed arise from the expression into the West. To most foreign observers, the origin of that spirit was obvious. It had nothing to do with the frontier family's encounter with the wilderness, and everything to do with its acquisition of landed property.

Ownership encouraged more than a sense of independence. Divided up into squares, the land that reminded the earliest colonists of God, and provoked a sensual hunger in later settlers, could also be treated by speculators simply as a commodity defined by numbers. A uniform, invariable shape that took no account of springs or hills or swamps was an obstacle to efficient agriculture, but to a financier tracking the rise and fall in land values, it was a great convenience. The grid, designed by Thomas Jefferson to create republican farmers, also turned out to be ideal for buying, trading, and speculating. The consequence was what D. W. Meinig, doyen of American geographers, termed "the most basic feature of the settlement process: That it tended to be suffused in speculation." The paradox was that most of the speculators were not big-time financiers—though there were plenty of *them*—but small-time republican farmers. Speculation, or, as the nineteenth century called it, capitalism, and democracy went together.

It took time for the movement to acquire momentum. In the first three decades of the century, sales rose from about 300,000 acres per year to around one million; but then the process suddenly accelerated. Between 1830 and 1837, more than fifty-seven million acres were sold. The General Land Office, which had been established in 1812 to supervise the survey, was almost overwhelmed, but the sheer size of the market was itself proof of the effectiveness of Jared Mansfield's squares. A sudden collapse in prices in 1837 and again twenty years later put momentary brakes on sales, and the discovery of gold in California in 1848 created a West Coast acquisition of land before the Midwest was half settled. Nevertheless, by the end of the nineteenth century more than a quarter of a billion acres of public domain had been converted into private property.

Caught by the romance of the frontier, one can easily miss what underpinned its spirit of restlessness, individualism, and enterprise. The adventure of taming the wilderness was certainly there, but what drew people from eastern states and from around the world was the desire for this soil magically transformed from wilderness to property by the act of measurement and mapping. The fact that the grid could fractally subdivide a continent into minute, graph-paper squares might appear to be simply a triumph of the mathematician's art, but the ease with which it made land available to anyone who went west in search of it had an almost incalculable influence on the development of the American psyche and the American economy.

THIRTEEN

The Shape of Cities

THERE WAS A SIMPLE TRUTH contained in the land survey: A shape that could be neatly reduced to smaller proportions was easy to sell. This had been apparent to John Jacob Astor as early as 1800 when he bought his first building on the Lower East Side of Manhattan.

There was something disgusting in Astor's inability to disguise his greed. He shoveled food into his greasy mouth, scooping up peas and ice cream with his knife, wiping his hands on the tablecloth—and he gobbled up land in Manhattan with the same unseemly haste, spending some $200,000 in one year alone to swell a hoard that eventually numbered more than 300 plots. His American Fur Company made a fortune trading beaver pelts in the far West, but he earned another simply by waiting for his property in and around the growing city of New York to rise in value. Land bought in 1800 for $50 an acre had a price tag of $1,500 in 1820, and when he died in 1848, Astor was worth $25 million, two and a half times as much as the next-wealthiest American.

Underpinning the rapid expansion of both Manhattan and Astor's wealth was the plan drawn up by the city commissioners in 1811 for New York's future growth. Most of the city was still concentrated into the southern tip of the island, below what is now Canal Street, although it was spreading northward, fueled by prosperity from trade through the port. As the three City commissioners explained in their report, they faced a crucial decision: "Whether we should confine ourselves to rectilinear and rectangular streets, or whether we should adopt some of those supposed improvements, by circles, ovals, and stars which certainly embellish a plan, whatever may be their effects as to convenience and utility."

The example of ovals and stars the commissioners were referring to was Washington, D.C., superbly laid out in 1792 by Pierre L'Enfant with the help of Andrew Ellicott and the African-American surveyor Benjamin Banneker; but in terms of development Washington, with its avenues radiating out from the Capitol and the White House, was a disaster. Despite the lure of being the nation's capital, there was no market in town lots there. It was Washington more than anything that finally broke the old speculator Robert Morris. Having acquired some 7,000 lots at $66.50 each, Morris counted

Figure 149. TThe Ellicott Plan for Washington, D.C.: 1792

Plan of Washington, D.C., 1792.

himself lucky to sell 500 of them, and ended up serving a six-year stretch in the debtors' prison in Prune Street, Philadelphia—which he airily called "the hotel with grated doors." When the duc de la Rochefoucauld passed through some years after the city's foundation, he remarked that Washington was no more than some buildings scattered among the woods, and that "most of them were built for speculation and remained empty."

Since one of New York's commissioners was Gouverneur Morris, Robert Morris's associate, it was certain that they would avoid Washington's example. "In considering that subject [the city's shape]," they wrote, "we could not but bear in mind that a city is to be composed principally of the habitations of men, and strait-sided and right angled houses are the most cheap to build and the most convenient to live in. The effect of these plain and simple reflections was decisive." Their choice was a gridiron consisting of a dozen north-south avenues crossed at right angles by 155 east-west streets. The distance between the the 100-foot-wide avenues was affected by the existing layout south of Houston Street, and consequently ranged from 610 feet to 920 feet, but there was nothing to constrain the planners in deciding the distance

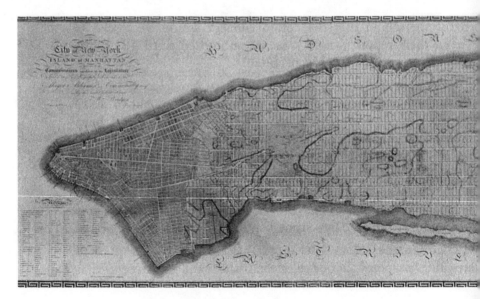

The commissioner's map of New York City, 1811.

between the streets. That was determined by the standby of every surveyor, Gunter's chain. As the grid moved into the farmland north of Houston (at that time North) Street, the depth of each block of Manhattan was going to measure 3 chains, or 198 feet. What this created was a thin, miserly block with an area of between 3 and 4 acres, compared to the more usual 5 acres in cities like Philadelphia and Chicago. Confined to so small an area, Manhattan developers would soon have no choice but to build upward.

The commissioners' solution resembled the land survey in ignoring natural features of the terrain, like hills and swamps, and in maximizing the opportunity for speculation. Once the plan was adopted, it became easy to predict where the city would spread. As Astor explained when a customer asked why he was selling a Wall Street location for $8,000 rather than holding on for a few years to get $12,000, "See what I intend doing with these $8,000. I shall buy eighty lots above Canal Street, and by the time your one lot is worth $12,000, my eighty lots will be worth $80,000." To criticism that the commissioners' plan was unimaginative, the city's surveyor, John Randel, would always point out that it was ideal for the "buying, selling and improving of real estate."

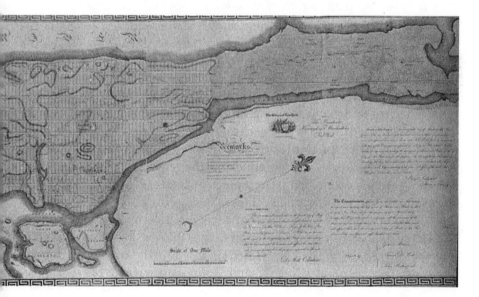

A process of simplification that decimalizers would have appreciated led the 198-foot depth of a block to be referred to as 200 feet, although as late as 1879 Frederick Law Olmsted, pioneer of urban parks and open spaces, complained bitterly that if a building "needs a space of ground more than sixty-six yards in extent from north to south, the system forbids that it shall be built in New York." These regulations gave rise to the standard Manhattan lot, generally described as a 100 feet deep (but in fact one foot shorter) and 25 feet broad, backing onto another of the same dimensions. Limited in area, developers compensated by cramming apartments into Randel's meager blocks. By the 1890's, a typical floor in a five-story tenement on the Lower East Side would be subdivided into four or more apartments, each costing around ten dollars a month rent, and housing ten or more people. According to one contemporary estimate, each acre housed about 1,000 people and as many as 2,800 in one Jewish tenement block. Epidemics of typhus and typhoid fever regularly ripped through what was at that date the most densely inhabited spot on earth, but the inhabitants' welfare attracted less attention than the fact that Manhattan's grid made it possible for one of its meager 3-acre blocks to generate an income of $12,000 a month in 1900, a time when $350 would keep a family for a year.

Manhattan's plain, profitable shape was to influence the design of American cities all the way to the Pacific, in contrast to Washington, whose intricate, unsalable pattern was rarely copied. (The magnificent heart of Indianapolis is the prime exception.) To twentieth-century town planners like Lewis Mumford, the gridiron was a disaster, mixing up residential and industrial areas, creating traffic problems and hideous slums. In John Reps's superb panorama of city design, *The Making of Urban America*, published in 1965, the Manhattan plan was depicted as little better than the plague: "The fact that it was this gridiron New York that served as a model for later cities was a disaster whose consequences have barely been mitigated by more modern city planners. . . . it stamped an identical brand of uniformity and mediocrity on American cities from coast to coast."*

*Since those words were written, a revolution in construction and design has transformed Manhattan. Yet while it may now be regarded as one of the architectural wonders of the world, its ground plan still causes snarl-ups, aids urban sprawl, and has helped create slums in cities across the nation.

In 1830 James Thompson, a surveyor and engineer, was commissioned to lay out a town in Illinois, in the square mile of Section 9, Township 39, Range 14, Second Principal Meridian, so that lots could be sold to finance the Illinois canal from Lake Erie to the Mississippi River. He took the Manhattan gridiron a step farther in regularity by dividing the section into squares measuring 20 chains by 20 chains—the homesteader's 40—and then subdividing each square into four blocks, whose dimensions of 10 chains by 10 included both buildings and streets. As had happened in Manhattan and across the land survey, speculators were attracted by the ease of multiplying or splitting up square and oblong properties.

Six years later, Harriet Martineau saw the results at first hand when she visited the mosquito-ridden swamp that Thompson named Chicago. "The streets were crowded with land speculators hurrying from one sale to another," she wrote. "A negro dressed up in scarlet, bearing a scarlet flag and riding a white horse, announced the time of sale. At every street corner where he stopped, the crowd gathered round him; and it seemed as if some prevalent mania infected the whole people." For the next seventy years, Chicago's grid spread outward in the undeviating pattern that Gunter's chain made so simple.

It was difficult for any urban plan to escape the influence, either direct or indirect, of the 22-yard length that was dividing up the country. Even in Thomas Holme's carefully conceived 1682 plan of Philadelphia, 10 chains by 10 is the space allotted to the five great open squares located at the city's center and four corners. It is noticeable that the street widths, carefully drawn in Holme's office, are 150 feet; but when Philadelphia's gridiron pattern was adapted for use in other cities, surveyors armed with chains found these decimal numbers hard to measure out on the ground. The solution was to "gunterize" them to 99 feet (150 links) and 49½ feet (75 links).

This was what happened when Joseph Smith, founder of the Church of Jesus Christ of the Latter Day Saints, designed a settlement in Jackson County, Missouri, in 1833. The square shape was based on the instructions for city-building in chapter 35 of the Bible's Book of Numbers, but the cubits in the original specifications were "gunterized" to rods and perches.

"We send by this mail a draft of the City of Zion," Smith told the Church.

Bird's-eye view of Salt Lake City, Utah, 1870.

"The plot contains one mile square; all the squares in the plot contain ten acres each, being forty rods [10 chains] square. . . . Each lot is four perches [1 chain] in front, and twenty [5 chains] back, making one-half acre in each lot." To maintain the town's regularity, all the streets were to be 2 chains broad. After Smith's murder by a mob in 1844, following his arrest on charges of treason and conspiracy, these dimensions were followed for each of the cities built by the Mormons in their search for a safe haven, the last and most spectacular of them being Salt Lake City, which Orson Pratt laid out on the directions of Smith's successor, Brigham Young, beneath the Wahsatch Mountains in Utah. Intriguingly, excavations for new sidewalks undertaken in Salt Lake City in 2001 revealed that the corner posts were set 4 inches farther out than they should have been, suggesting that Pratt had not followed the practice of public lands surveyors and calibrated his chain precisely before starting work.

*T*HERE WAS NOTHING INEVITABLE about these urban shapes. In New England the colonists had developed the village pattern, in which the original purchasers settled around a green at the center of the community, with fields behind their houses, and grazing land still farther out. In Montreal, Mobile, and New Orleans, the French planners produced designs centered on a hollow square from which narrow streets ran out, lined by houses with gardens and courtyards behind. In the province of Quebec, they experimented with fortified villages at the center of a star of wedged-shaped fields. The impact of the 1526 Law of the Indies, which did envision a grid for settlements in Spanish America, was limited in its effect to the streets around the central square, and beyond that the typical pueblo usually developed in haphazard fashion, more in response to the availability of water than to planners' dictates.

Once the U.S. land survey established its pattern, however, it was hard to resist its rectangularity. The chain's length was so well fitted to the section's 640 acres that few city planners could resist incorporating its dimensions into their designs, especially when speed and quick sales were required. Even the radiating design of Indianapolis had to fit inside a section's one-mile-square box, and the sad story of Circleville, Ohio, which was planned as a series of rings but was speedily converted to squares by the residents, proved how hard it was to break the pattern.

*T*HE RAILROADS MADE THE MOST extreme use of the grid and the chain. As an incentive to build, the federal government awarded railroad companies blocks of land, usually a township but sometimes more, alternating on either side of the track. Most of it was sold off to farmers, but the real money lay in towns. Companies like the Northern Pacific and the Burlington designed a standard town that could be laid out and sold off wherever they decided to site a passenger and freight depot. The basic model consisted of three 160-acre sections on each side of the track, each section being split four ways into those 40-acre—20 by 20 chains—lots that a surveyor could measure with his eyes closed.

"An office boy can figure out the number of square feet involved in a street opening," Lewis Mumford exclaimed, "and a lawyer's clerk can write a description of the necessary deed of sale merely by copying a standard document. With a T-square and a triangle, finally, the municipal engineer without the slightest training as either an architect or a sociologist could 'plan' a metropolis."

This of course was the virtue of the scheme in the railroad companies' eyes. Seven dollars was as much as the Burlington was prepared to pay to

Bird's-eye view of Herington, Kansas, 1887.

have a town planned. Sometimes the companies built only on one side of the tracks, in which case they simply cut the plan in half, but the depot always remained the focal point. In the West a depot was required every 12 or 15 miles so that farms would not be more than half a day's journey away, and hundreds of identical towns were laid out with a railroad down the center, two parallel east-west streets on one side named Oak and Chestnut, and on the other side two more named Walnut and Hickory, crossed by ten others named First to Tenth. There were so many that it was hard to come up with names for them all.

"I shall have two or three more towns to name very soon," Charlie Perkins, land agent to the Burlington Railroad, wrote in the 1870s as the track advanced through Iowa. "They should be short and easily pronounced. Frederic I think is a very good name. It is now literally a cornfield, so I cannot have it surveyed, but yesterday a man came to arrange to put a hotel there. This is a great country for hotels."

The seven-dollar plans were hyped in beguiling fashion. "Would you make money easy?" asked George F. Train, a founder of the Union Pacific Railroad. "Find then the site of a city and buy the farm it is to be built on! How many regret the non-purchase of that lot in New York; that block in Buffalo; that acre in Chicago, that quarter-section in Omaha." This was the more honest approach, though Train himself died a pauper. The less honest but more common method was to sell the farmland and pretend that the towns were already built and thriving there. "*Towns* they are on paper," confessed the amiable Charlie Perkins, "meadows or timberland with here and there a house, in reality."

Not until the beginning of the twentieth century did the planners succeed in fighting free of the sort of drab materialism the square encouraged. The turning point was the 1893 Chicago World's Fair, where examples of spectacular and innovative architecture triggered such enthusiasm that Daniel Burnham, the founder of American town planning, was able to persuade the city authorities in Washington, Cleveland, and Chicago to redesign public buildings in his trademark style of gleaming white stucco, decorated with domes and pillars and set off by parks and esplanades. When San Francisco was all but destroyed by earthquake followed by fire in April 1906,

Bird's-eye view of San Francisco, California, 1868.

Burnham produced a new city plan with more parks, curving roads, and diagonal thoroughfares to take the place of the plain grid hurriedly laid down in the days of the gold rush. As cities grew, suburban planners followed the advice given by Frederick Law Olmsted that they should adopt "in the design of your roads, gracefully curved lines, generous spaces, and the absence of sharp corners, the idea being to suggest and imply leisure, contemplativeness and happy tranquillity."

Yet even as this idea took hold, and developers began to create the concept of suburban living that deliberately shunned the sharp right angles of farms and city blocks, an old reality still lurked below the surface. The original real estate was still being bought by the square 40-acre lot, house plots were sold in 10-, 5-, or, most commonly, 2.5-acre parcels, and streets tended to measure 66 feet, or 2 chains, in width. As one Illinois developer observed in 1966, "Underneath all these contemporary trappings, our basic thinking is still geared to a gridiron block system."

That system's resilience was apparent, too, when San Francisco arose from the ashes of earthquake and fire. Instead of following Burnham's graceful curves and diagonals, the city kept instead to the old gold-rush grid that climbed straight up Nob Hill and hurtled straight down to San Francisco Bay. Even John Reps was prepared to admit that for once squares and straight lines could be exhilarating. "San Francisco is a glorious exception to the otherwise gloomy record of the grid," he wrote. "On no site less superb could this have happened, nor is this, visually the greatest of America's cities, likely to be duplicated."

Jasper O'Farrell, who laid out San Francisco's original grid in 1845, would have been gratified, but he would have noted, as well, that the new plan designated the dimensions of a city block in feet and inches. The unit he had used to measure it out was the Mexican *vara*. Since then, a minor miracle had occurred. The American Customary System of Weights and Measures had become the law of the land.

FOURTEEN

Hassler's Passion

IN DECEMBER 1806 Thomas Jefferson's former ally in the reform of weights and measures, Professor Robert Patterson of the University of Pennsylvania, sent a letter to the president recommending the employment of a Swiss immigrant named Ferdinand Rudolph Hassler. "In his education he paid particular attention to the study of astronomy and surveying," Patterson wrote. "He is a man of a sound, hardy constitution, about thirty-five years of age, and of the most amiable, conciliating manners."

Even this glowing recommendation did not convey Hassler's unique quality. He was a geodesist, an earth measurer. As a student at Bern University, he had met Johan George Tralles, a gifted scientist and father figure, who introduced him to the exquisite pleasure of land measurement coupled with the harsh discipline of pinpoint accuracy, and the masochistic science captured Hassler for life. In 1791 he and Tralles undertook the first geodetic survey of the canton of Bern in his native country. The report of the Economic Society of Bern that led to the project emphasized the economic value of knowing the precise heights of hills and valleys, and the exact location of natural and man-made features, when planning major engineering works like roads and canals. But the difference between a geodetic survey and any other kind was not just a matter of economic advantage. A geodetic survey was so precise that it increased the supply of the world's knowledge about

Ferdinand Hassler

the shape of the earth and, in the words of the report, would "secure the thanks of the world of learning and be a lasting honour." It was an early example of the alliance of science and industry that would mark the nineteenth century.

Two years later Hassler and Tralles traveled to Paris, where they met Delambre, Lavoisier, and other leading figures of the metric commission and acquired copies of the meter and the kilogram. Both men became enthusiasts for what Hassler termed "the best and most consistent system of weights and measures hitherto devised," and in 1799 Tralles would take a leading role in the international commission that verified the scientific basis of the metric system.

On his return to Switzerland Hassler fell in love with Marianne Gaillard, whose surname, "gaiety," conveyed her lighthearted nature. She had been brought up to sing and play the piano, and to occupy the role of wife to a comfortable Swiss public servant. Her husband's personality, however, nursed a contradiction exhibited by so many land measurers that it is almost a defining characteristic—a passion both for exact definition and for untamed wilderness. When in 1803 the French took over the triangulation of Switzerland on which Hassler had been employed, the buried side of his nature came to the surface.

In 1804 he decided to emigrate and start a colony in the Southeast of the United States. On arrival in Philadelphia, the Hasslers discovered that the agent to whom they had entrusted their money had gambled it all on a fraudulent land deal and lost everything. Poverty forced Ferdinand to stay in the city rather than going off to be a farmer in the country. Almost at once he became involved with the American Philosophical Society, of which both Patterson and Jefferson were members.

Among the objects packed into the ninety-six trunks and boxes Hassler brought with him were his kilogram and meter. To keep the family fed, he was forced to sell these precious items and most of his instruments to John Vaughan, a member of the Philosophical Society; but his expertise in French decimal measurement was noted, and some metric enthusiasts asked him to survey a long enough stretch of the Hudson River valley to provide an American basis for the meter. The project came to nothing, but it persuaded several other members of the Philosophical Society to write to Jefferson about the remarkable Swiss scientist in their midst.

These letters had a remarkable outcome, for they finally enabled Jefferson to achieve the decimal measurement of the United States that he had dreamed of since 1784. In 1807, the last year of his presidency, Jefferson persuaded Congress to authorize funding for the United States Coast Survey. As its name implies, the survey was intended to provide accurate maps of the eastern seaboard. They were needed for defense and for the huge maritime traffic that carried America's goods between the Mississippi Valley and the Atlantic coast, and onward to the outside world—but these were merely the practical reasons. Together with Albert Gallatin, his secretary of the Treasury, who had a grounding in mathematics and a commitment to science as firm as his own, Jefferson intended to give the scheme a wider context.

The geodetic survey of Switzerland was only part of an international trend to precision mapping. France led the way and by the end of the eighteenth century had been measured from top to bottom three times. When Delambre presented Napoleon with his acerbic memoir of the final, meridian survey, *Base du système metrique décimal* (Basis of the decimal metric system), the scale of the enterprise left even the emperor awed. "Conquest is temporary," he said modestly, "this work will endure."

Across the Channel the Ordnance Survey was in the midst of its seventy-year project of triangulating Great Britain. In the Netherlands, Scandinavia, Russia, and the Austro-Hungarian Empire, governments had started or were about to embark on similarly exact and scientific surveys of their territory, using triangulation to calculate distance, celestial observation to establish location, and barometric pressure to indicate height. From the Arctic to the Mediterranean, from Ireland to the Urals, a series of geodetic surveys, precise enough to establish where each town and mountain stood on the earth's surface, were beginning to cover Europe.

By contrast, much of the United States remained unexplored, and maps of the country were still fragmentary and small-scale. The only official, large-scale record was the series of surveyors' plats compiled at speed using chain and compass, and their technical name, "cadastral" or "property registering," indicated their limited purpose. It was this lack that Jefferson and Gallatin intended the Coast Survey to fill. Whatever Congress might have had in mind, they wanted a scientific agency, the first created by the federal government, which would do for the United States what the Cassini family and Jean-Charles Borda had done for France, and William Roy and the Ordnance Survey were doing for Britain.

It seems clear that they also knew whom they wanted in charge. Despite shortcomings in both language and temperament—toward the end of his life the New York newspaper *New World* referred to him as "an old Swiss named Hassler who writes a miserable jargon which he calls English and scolds like a fish hag"—Hassler's proposal for the survey, written in French, was selected as the model to be followed.

The proposition that he put to Gallatin was to measure the Atlantic coast with a chain of enormous triangles whose sides would be 60,000 to 100,000 feet long, resting on two precisely measured baselines. In the "miserable jargon" he called English, he emphasized the utter necessity of using only the finest instruments, including telescopes, theodolites, barometers, thermometers, and chronometers, explaining that "good instruments are never to be found in shops, where only instruments of inferior quality are put up for sale. They must be made on command, and by the best mechanicians of London."

The supremacy of British instrument makers was universally acknowledged. At that moment, Benjamin Latrobe, a surveyor with the Chesapeake and Delaware Canal Company, was writing to his employers about a telescope level he had bought from a Mr. Biggs, Philadelphia's finest maker of "nautical, optical, geographical, surveying, gauging, gunnery, drafting, drawing, and leveling instruments." It was, Latrobe commented sniffily, "truly made for the American market as to filing, gilding, and engraving. It could not have been sold, I believe, in London."

Only Britain had the metallurgists and machine tools, the glassblowers, chemists, and metalworkers, to produce instruments of the required accuracy. The most exacting craftsmen—"artists," Hassler called them—were to be found in London, and the best of them was Edward Troughton. The German astronomer H. C. Schumacher once waited four years to get a zenith sector made by Troughton because no other was good enough. For the scientist Sir George Shuckburgh, Troughton constructed a yard with a built-in micrometer attached to a microscope so that the measurements marked in hairlines could be examined to an accuracy of one-thousandth of an inch. The master of the Mint and General William Roy both ordered their definitive standard yards from Troughton; but Hassler paid him the finest compliment of all, naming his newborn son Edward Troughton Hassler.

Unfortunately for Jefferson's plans and Hassler's career, a trade embargo and then the War of 1812 with Great Britain delayed the acquisition of these essential instruments, and almost ten years passed before the Coast Survey began, by which time Jefferson had been succeeded as president by James Madison. A small income from teaching, first at West Point and then at Union College in Schenectady, New York, kept Hassler's family from starving, but it was not until July 1816 that he laid out the first two baselines with infinite care, along the beach of what is now Coney Island and at Englewood, New Jersey. Despite the fineness of Hassler's instruments and the exactness with which he measured the azimuth of the Polestar, any chainman on the land survey would have been puzzled by the sight of his measuring chain. It had nothing to do with Gunter. Instead, it consisted of eight links, each of which was exactly 1 meter long. Its length was measured against the meter that Hassler's friend Tralles obtained for him in 1799 while a member of the international metric commission.

After just one summer's work Congress closed down the survey for lack of money. Hassler retreated to Long Island, where he tried being the farmer he had originally wanted to be; but Marianne could not stand the isolation and ran away. Most of his earnings came from writing textbooks, and it was not until 1833 that the Coast Survey was restored, with Hassler hired again as superintendent. He immediately celebrated with another London buying splurge: a new 30-inch repeating theodolite built to his own design by

Troughton, two microscopes from Dollond, and a Ramsden dividing engine so that in future the United States could make its own precision instruments. This time he used four iron bars, each 2 meters long, laid end to end to measure the baseline, and his triangles were so meticulously calculated that when they were checked in the 1970s the error rate was found to be just one in 100,000.

Next to prickly Rufus Putnam, the Swissly precise Hassler is one of the most appealing measurers of the United States. Nothing distracted him from his task. In summer and winter, he dressed for simplicity's sake always in loose white flannel, merely subtracting or adding layers as the temperature rose or fell. Despite his strange accent and hopelessly unmilitary appearance, his West Point students adored him, and almost forty years later one of them, the redoubtable Colonel Joseph Swift, commander of the Corps of Engineers, wrote to Hassler's family, recalling affectionately "the intimacy between your father and myself . . . that remained unbroken through Mr. Hassler's undeserved vicissitudes to his death."

To understand Ferdinand Rudolph Hassler, so odd in behavior, so exact in performance, is to appreciate cuckoo clocks. For his array of heavy brass instruments—the Troughton theodolite alone weighed 300 pounds—he had a special carriage built with extra-wide wheels, heavily braced springs, and cushioned boxes. It was painted canary yellow and contained a suspended desk so that it could act as an office, and a folding bunk with a locker below holding wine, biscuits, and cheese so that it could act as a home. When he grew shortsighted in old age, Hassler refused to be bothered with glasses but took to flicking snuff in his eyes—"To excite the optic nerve," he explained—so that as he bent over a map the stinging tears acted as a temporary contact lens. Nothing was allowed to distract him from the incessant drive for accuracy.

It was Hassler, for example, who discovered that most meter bars were minutely shorter than they should be, because while the ends were being filed off to bring them to the correct length the metal heated slightly and expanded. Once it cooled, what had been a meter contracted by something like six-hundredths of a millimeter. Hassler's concern for these tiny quantities was a symptom of his modernity. When America's two leading surveyors of the eighteenth century, Andrew Ellicott and Simeon de Witt, New York's surveyor-general, worked with him on defining the frontier with Canada, they found his approach frustrating. Writing to a friend, Ellicott complained that "not more than one observation in ten can possibly be applied to the boundary—those that can are probably good, but their mode of calculation is laborious in the extreme." But Hassler was not going to lower his standards for anyone. When at last he was allowed to begin work on the Coast Survey, he spent forty-three days measuring a baseline less than 9 miles long.

This meticulous insistence on accuracy drove Congress to distraction, and messages winging from Washington to the yellow carriage would demand to know how long it would be before Hassler finished. In 1843 the *New World* declared that it was "an evil hour when [Hassler was] selected as chief surveyor. He has been engaged in this for 13 [*sic*] years at the rate of 6000 dollars per annum, and as the results are not forthcoming Congress naturally wishes to know what he is about." That winter he died, his triangles having reached no farther than the southern border of New Jersey. But by then the exacting standards he set had become part of the United States Coast Survey. As a result, it took another fifty-five years to survey the entire coastline from Maine to New Orleans, and every yard of it was measured in meters. Later the Coast Survey was extended to cover the entire United States. The words *and Geodetic* were added to its title, and the whole landmass was mapped in the same careful, metric fashion.

*T*HERE DOES NOT SEEM to have been any dissent at the time to Hassler's decision to use the metric system. The only measure specified by Congress was the league—it limited the survey's scope to "within twenty leagues" of the coast—and he would not have been expected to use that archaic unit. His 1807 proposal for triangulating the coast expressed distance in thousands of feet. Yet there was never any question in his mind about the superiority of the metric system. Not only was it more convenient, he once wrote, it "produced results of considerable utility in many other respects, improvements in mathematical and natural sciences, and in mechanical arts, besides the establishment of the best and most consistent system of weights and measures hitherto devised." For a scientific project like a geodetic survey, no other system was conceivable.

Jefferson, having left office before the survey began, does not appear to have been aware of Hassler's innovation. He never warmed to the meter, regarding the meridian basis as an unnecessary act of French nationalism which, as he explained in a letter to Patterson, forced other nations "to take their measures from the standard prepared by France." But his enthusiasm for decimals did not diminish—one of the gadgets of his old age was an odometer that divided the mile into hundredths—and he boasted, "I find every one comprehends a distance readily when stated to them in miles & cents; so they would in feet and cents, pounds & cents, &c." He had always supposed that once decimal measurement was adopted by "the men of science," it would only be a matter of time before it was taken up by "the tardy will of government who are always in their stock of information a century or two behind the intelligent part of mankind."

John Quincy Adams

That was also the view held by John Quincy Adams, the son of Jefferson's old antagonist and friend John Adams. In 1821, as secretary of state, he delivered a report to Congress on the feasibility of "establishing uniformity in weights and measures" in response to a request from the Senate. He made it clear that whatever changes were made would happen very slowly indeed, and illustrated his conclusion with a wry look at the lack of success of the nation's decimal coinage in displacing the old British and Spanish currencies.

"Even now at the end of thirty years," Adams wrote,

ask a tradesman or shopkeeper in any of our cities what is a dime or a mille, and the chances are four in five that he will not understand your question. But go to New York and offer in payment the Spanish coin, the unit of the

Spanish piece of eight [an eighth of a dollar], and the shop or market-man will take it for a shilling. Carry it to Boston or Richmond, and you shall be told it is not a shilling but nine pence. Bring it to Philadelphia, Baltimore or the City of Washington, and you shall find it recognised for an eleven-penny bit; and if you ask how that can be, you shall learn that, the dollar being [equivalent to] ninety pence, the eighth part of it is nearer to eleven than to any other number: and pursuing still farther the arithmetic of popular denominations, you will find that half eleven is five, or at least that half the even-penny bit is the fi-penny bit, which fi-penny bit at Richmond shrinks to four pence half-penny, and at New York swells to six pence. And thus we have English denominations most absurdly and diversely applied to Spanish coins; while our own lawfully established dime and mille remain, to the great mass of the people, among the hidden mysteries of political economy— state secrets.

Had he been able to look farther ahead, Adams might not have been astonished by an official report made by a State Department representative to an international conference in Berlin in 1862, which revealed that almost eighty years after the decimal dollar became the currency of the United States, "shopkeepers still more readily say two shillings and sixpence than thirty-seven and a half cents." But even he might have been startled to find that in the twenty-first century, the quarter was still known as two bits, and that the New York Stock Exchange continued to express stock values in eighths and sixteenths of a dollar, as though Jefferson's decimal version had never supplanted the old Spanish kind.

Taking the long view, Adams was not inclined to recommend any urgent changes. According to his report, the existing system was chaotic but he could not see any practical advantage in changing to a decimal system. "A glance of the eye is sufficient to divide material substances into successive half, fourth, eighth and sixteenth. . . . But divisions of fifth and tenth parts are among the most difficult that can be performed without the aid of calculation."

Nor did he think that the metric system's scientific basis was especially relevant. "It is of little consequence to the farmer who needs a measure for his corn, to the mechanic who builds a house, or to the townsman who buys a pound of meat or a bottle of wine to know that the weight or the measure which he employs was standardized by the circumference of the globe. Should the meter be substituted as the standard for our weights and measures instead of the foot and inch, the natural standard which every man carries with him in his own person would be taken away." His conclusion was that of opponents of the metric system for centuries to come: "The convenience of decimal arithmetic is in its nature merely a convenience of calculation."

In the long term, Adams recommended that Congress should consult with other nations about cooperating with France to "the final and universal establishment of her system," but meanwhile the best the United States could do was to wait until Britain, the economic giant of the age and the nation's closest trading partner, had established standards for its own weights and measures. Congress needed no other excuse to do nothing.

But Adams was wrong. Inactivity was no longer an option. The confusion of measures used in the thirteen original states had been carried by the public lands survey into ten new western states. On being admitted to the Union, most had legislated that standards of "English measures" were to be used, without distinguishing between the different sizes of bushel, barrel, and gallon. In Indiana and Illinois, for example, Virginia measures were used in the south of the state and New England units in the north. Vermont and the District of Columbia had inherited their own private confusions from their most immediate neighbors.

Louisiana, where 50,000 French inhabitants had been converted by the Louisiana Purchase into Americans, faced the most contentious situation. They had always defined weights by the *livre* and *quintal*, and lengths by the *pied* and the *toise*; their rice was measured by the *boisseau*, their land by the *arpent carré*, and their cloth by the *aune*. Since 1814 it had been illegal to use any but American measures (in practice Georgia's), with fines of fifty dollars for offenders, but when Monsieur Bouchon, Louisiana's Surveyor-General, wrote to Adams in 1820, he could report only mixed success. "The ancient inhabitants are well enough satisfied with the American weights," he noted, "but all, and especially those in the country, find it very difficult to accustom themselves to the measures of length and [of area] and I think they will be long in doing so."

This was a shrewd prediction. For years the characteristic shape of a Louisiana farm remained a narrow strip measuring 200 square *arpents* (about 166 acres), and the resistance of farmers to the acre forced land surveyors to abandon their squares around New Orleans in favor of the strip. (To this day, some Louisianans still refer to real estate in *arpents*.) But as might have been expected, in New Orleans itself traders were more flexible, taking advantage of the change to buy goods by the *pied*, which was almost 13 inches, and sell by the 12-inch foot.

Confusion by itself was not enough, however, to break the congressional repose. It was the trade between these western states and the East that finally forced them to act.

*I*N 1825 Governor DeWitt Clinton of New York opened the 350-mile-long Erie Canal linking the lake with the Hudson River, and predicted that, "as an organ of communication between the Hudson, the Mississippi and the St. Lawrence, the great lakes of north and west and their tributary rivers, it will create the greatest inland trade ever witnessed." That turned out to be a modest forecast, for a ton of corn that had taken three weeks to be carried by wagon from Buffalo to New York at a cost of $100 could now be put on a barge and for just $6 would be delivered eight days later. Suddenly, the produce of the Midwest—millions of tons a year of grain, timber, and coal—that had been carried down the Mississippi or painfully by mule and wagon across the Alleghenies, gushed instead through the canal into the port of New York.

The Erie Canal transformed New York into the premier market of the United States, and just as British tobacco merchants had imposed their own definition of a hogshead on Virginia tobacco growers, so New York brokers buying from Midwest farmers insisted on their own tons and bushels. But a New York bushel might contain as many as 2,175 cubic inches of wheat or as few as 2,104 cubic inches. The first figure applied to purchases and the second to sales; multiplied by several hundred thousand tons a year, the disparity contributed significantly to the profitability of New York's trade.

Midwest farmers suffered, but so did the United States's revenues, the greater part of which came from tariffs levied on goods passing through New

York. Few customshouses had standards of their own—in New York, federal customs officers depended entirely on the state's measures—and it was a poor trader who could not avoid paying at least some of his dues. The new states were above all farming states, whether they produced corn, wheat, or cotton. It was in their interest not only to have uniform weights and measures but to have revenue raised from tariffs rather than land taxes.

It was, therefore, no coincidence that when the United States started to establish an official set of measures, the president was Tennessee-raised Andrew Jackson, the personification of the new western outlook. Nor did chance have anything to do with Jackson's choice of a commissioner to examine the state of measures held by U.S. customshouses. There was only one person in the country whose obsession with accuracy to the last observable fraction of a measurement matched the task. In 1830 Ferdinand Rudolph Hassler was appointed to undertake the examination that was to lead the United States to a uniform system of weights and measures.

At the age of sixty, Hassler's life was a failure. The Coast Survey had been stopped; his wife had left him; he had lost all his money, sold his survey instruments, his unique collection of metric and organic measures, and his library of 3,000 volumes; and he survived by writing textbooks and eating cheap meals in Kinchy's Italian restaurant in New York. He had grown rather solitary, certainly a little cranky, although it was hard to tell whether this was a change or an exaggeration of something already present.

While they were together, he and Marianne had produced nine children, whom Hassler clearly loved. He made their clothes, baked bread and pastries for them, and as adults two of his sons chose to work with him. Nevertheless, as a husband he must have been exasperating, alternately absentminded and stubborn about doing things exactly right. His capacity for concentration had always been extreme, making him, as one member of his class at Union College recalled, a hopeless teacher: "He became so absorbed in demonstrations, his students would walk out on him, and he would never be conscious of their absence." They made up a story about him reading a letter and being so lost in its contents that he tripped and fell, yet continued reading while he lay on the ground until he had finished it. To those who knew Hassler, it sounded credible.

Despite his strange appearance—the red-rimmed, snuff-inflamed eyes, the white suit stained with wine and crumbly with biscuits—Hassler was not a freak. In his single-minded passion for accuracy, he was simply a man before his time. A generation later, science had adapted itself to the supreme importance of precision, writing up every detail of how experiments were conducted, and recording results to three and four places of decimals. In 1832, when Hassler produced his report on the customshouse measures for President Jackson, he still had to make the case for exactness, and in a passage as remarkable for its passion as for its fractured English he delivered what was in effect his testament of faith: "A system of weights and measures must be a careful scientific operation in which the accuracy aimed at must far exceed that required for practical use. . . . What is done in such a work is done for the future, the improvement of science always spreading more in common life; if such [a standard of accuracy] is not ahead of its time, even if possible of the science of it, it very soon drops back, behind even the wants of the nicer social intercourse; and such an epoch approaches always more rapidly with the greater means of science." Which meant, roughly, that science brought progress, that the better the science the faster the progress, but that progress must inevitably consume the science that made it possible.

Yet Hassler did not value accuracy for its own sake, or even for science's, but, as the opening words of his report explained, because "among the means of distributing justice in a country, which is the aim of the establishment of Governments, rank unavoidably the fixation and distribution of accurate weights and measures for all the daily dealings of active life."

Although he had introduced the metric system to the Coast Survey, Hassler recognized that it had become impossible to do the same for the United States as a whole. But as a believer in the system, he could not help expressing his regret that Jefferson had failed in his efforts to link reform of weights and measures to his design for creating states out of the empty west. "It would undoubtedly have been a great advantage to the country," he wrote, "had a regular system, founded upon a scientific base, and single unit been established together with the first regulations for the organization of the country, before the great increase of population, and consequent great active intercourse had created and increased the difficulty of attacking old habits."

That moment had passed. In 1790 there had been fewer than five million inhabitants to convince of the need for change; in 1830 there were almost thirteen million, nearly all of them used to traditional, four-based measures. Hassler's task was simply to establish standards of uniformity, and some of the work had already been done for him.

In 1824 the British government had at last replaced all the conflicting measures that so upset the Carysfort committee eighty-five years earlier and established a single system not only for itself but for the empire. The imperial weights and measures were based on three fundamental standards: Bird's 1760 yard, which itself was derived from Elizabeth I's 1588 measure; Bird's 1758 troy pound, which was taken from the gold merchants' and apothecaries' measure made legal for the first time in Henry VII's 1496 statute; and a single gallon size for both liquid and dry measures, defined as being large enough to contain 10 pounds avoirdupois of distilled water. The first two were utterly traditional, but the gallon (equivalent to 277.421 cubic inches) was an innovation born from the committee's inability, even after ten years of intellectual bloodshed, to agree on one definitive version out of all the different, contradictory ale gallons, wine gallons, and corn gallons already on the statute books.

Having battled their way to agreement on the standards, the members of the measures committee saw their work destroyed in 1834 when the Houses of Parliament, where the Imperial yard and pound were kept, caught fire, and in the heat the two meticulously engineered originals became bent and distorted. Another ten years passed before replicas of sufficient accuracy could be constructed, but from these two units were derived all the others, from grains to tons, and from inches to miles. Hundreds of standards were sent out to local authorities, and magistrates were ordered to enforce their use.

One of the units, the troy pound, had already become an unofficial standard for the United States after a brass copy was purchased by Albert Gallatin while ambassador in London, and presented to the Philadelphia Mint in 1828. It remained the official standard for measuring the weight of U.S. coins until 1911. Since an avoirdupois pound could be derived from the troy variety, this smooth, circular brass weight served for a time as the absolute standard for all American weights. The presumption was that Hassler would take other imperial units to serve as his standards.

Hassler's report took more than a year to compile. Many of his experiments were carried out in winter so that instruments could be tested above and below the freezing point. For standards he used the troy pound in the Mint and the splendid brass 82 inch Troughton measure he had had made for the Coast Survey in 1814. The yard and every gradation from 0 to 80 inches were cut into it. But to satisfy Hassler's lust for exactness, John Vaughan, who had bought the Tralles meter bar and kilogram so that Hassler had enough money to eat, lent them back to serve as additional checks.

In a typical passage in his report, Hassler explained the difficulty of comparing a customshouse weight with one of his standards: Each had to be placed in a scale whose point of balance was for utmost sensitivity a knife-edge, but its accuracy "depends on the sharpness of the knife edges, and these must lose their edge under a heavy pressure; thence besides their very expensive construction, they require the constant presence of an artist to repair them after a few weighing operations." There was always a price to be paid for Hassler's work, but no one ever doubted that the knife-edge of his scales would cut like a razor. When it proved impossible to maintain that sharpness, he built a hydrometer in which weight was determined by the displacement of water.

Hassler's report revealed what the secretary of the Treasury Louis McLane, called "a serious evil," although he was referring to loss of revenue rather than the injustices that resulted from different measures being used, or in some cases being totally absent, in ports from Boston to New Orleans. "It is believed, however," McLane continued, in a sentence that was to signal the breaking of the logjam, "that this department has full authority to correct the evil, by causing uniform and accurate weights and measures, and authentic standards to be supplied to all custom-houses."

The standards were to be made at the U.S. Arsenal in Washington, "under the immediate personal superintendence of Mr. Hassler . . . with all the exactness that the present advanced state of science and the arts will afford." If that rang like the sound of trumpets in Hassler's ears, it must have seemed as though the full orchestra had struck up in August of that year, when the Coast Survey was reinstituted. For the last ten years of his life, he remained in charge of both, working on the survey in summer and the weights in winter.

In constructing the standards, Hassler displayed the full range of his gifts for exactness. Copies of weights from around the world were ordered, alloys of metal were assayed, then weighed in a vacuum and at normal sea-level pressure, at freezing and at sixty-two degrees Fahrenheit. But that was only the start. Eventually, each standard had to be weighed on precisely calibrated scales. Because the tiniest movement of air would affect the scales, Hassler covered up the windows of the laboratory to prevent drafts and to stop the sun's rays from heating the room and creating temperature currents. "I took besides the precaution to nail to the side of the bench directed toward the windows, papers that reached several inches above it to prevent any draft over the scale," he reported, "and other large whole sheets [were nailed] on the side toward the observer, near the microscopes, so as to intersect the communication of the heat of the body of the observer to the microscopes and scale, or the effect of his breathing." Having set up the laboratory, Hassler would leave everything in place until the following day so that the disturbed air and temperature had the opportunity to settle down. On his return he wore gloves so that when he placed the weights on the scales no heat would be transferred from his hands.

As reports on Hassler's construction of the customshouse standards reached Congress, his extraordinary commitment to the uttermost achievable degree of accuracy began to affect the way congressmen thought about his goal. It was still pictured more in terms of revenue than of justice, and Hassler was regarded with irritation rather than respect, but for the first time since George Washington had reminded Congress of its constitutional responsibility for weights and measures, the subject was taken seriously. On June 14, 1836, a joint resolution of the Senate and the House of Representatives ordered that "a complete set of all weights and measures adopted as standards [for customshouses] . . . be delivered to the governor of each State in the Union . . . to the end that a uniform standard of weights and measures may be established throughout the United States."

What was odd about this, and remains astonishing, was that there had never been any formal definition of what those weights and measures should be. It was simply left to the secretary of the Treasury, who took his lead from Hassler—if it was good enough for him, it was good enough for the United

The yard within Troughton's 82-inch scale.

States. And so from making standards for the customshouses, Hassler went on to make a set for every state in the Union. It was not until 1857 that the weights and measures he made were formally made legal throughout the land as the American Customary System of Weights and Measures.

The yard was specified as the distance between the 27th and the 63rd inch engraved on a silver scale inlaid down the center of Troughton's 82-inch scale, which is still preserved in the museum of the National Institute of Standards and Technology in Washington. The basic weight was the troy pound in the Philadelphia Mint. This was scaled up by the ratio of 5,760 grains to 7,000 grains to give the avoirdupois pound, the actual standard being a brass weight marked with a star, and consequently known as the "star" pound. Both of these were the same as the British imperial units, but instead of selecting the new all-in-one British gallon, the U.S. Treasury decided that the distinction between liquid and dry measures should be kept, initially for customs duties and then for the standards themselves. From the grab bag of old definitions, Hassler's elaborate comparisons of different measures determined the selection of Queen Anne's wine gallon of 231 cubic inches for liquids, and for grains the Winchester bushel of 2,150.42 cubic inches, each of which was smaller by about 17 per cent and 3 percent, respectively, than the new imperial gallon and bushel.*

---

*The connection between capacities and weights was expressed informally in the phrase "A pint is a pound the whole world round."

The "star" pound.

For convenience, he recommended no standard for the British stone, weighing 14 pounds, which quickly dropped out of use in the United States, and decided to clean up the incongruity of a hundredweight weighing 112 pounds. In the United States it would weigh 100 pounds, and the U.S. ton would come in at 2,000 pounds rather than the British 2,240 pounds. It was a gesture toward decimalization.

Since the pound was Roman in origin, and the yard Saxon, while the wine gallon was first recorded in the Magna Carta, and the Winchester measures could be traced back to King Edgar in the tenth century, the set of weights and measures that emerged from Hassler's workshops was truly traditional. In science he preferred the stark simplicity of the meter, but for practical purposes he recognized that the choice "of a set of standards in general depends upon the individual use made of them," and for the United States that meant the old four-based measures. It was his unique achievement to have established both systems for use in his adopted country.

In November 1843 Hassler took a hard fall while trying to protect one of his precious surveying instruments in a storm, and a few days later he caught pneumonia and died. As often happens, all the stories of his exasperating behavior instantly became affectionate reminiscences. One, told by President Jackson, was of Hassler coming to demand that his salary for the Coastal Survey be raised to $6,000 a year. When Jackson objected that this was as much as Hassler's boss, the secretary of the Treasury, was being paid, Hassler exclaimed in his uncertain but utterly truthful English, "Plenty Mr. Everybodys for Secretary of ze Treasury, but—only vone Mr. Hassler for ze Head of ze Coast Survey!" Recognizing that in this, as in everything, Hassler's accuracy could not be questioned, Jackson gave him the raise.

FIFTEEN

The Dispossessed

AN ESTIMATED 70,000 PEOPLE passed through the Crystal Palace in London on May 1, 1851. Everything about Sir Joseph Paxton's structure was fabulous: Walls and arched ceilings of clear glass prefabricated around an intricate skeleton of slender cast-iron rods soared more than 100 feet in height, stretched more than 600 yards in length, and covered an area of 23 acres. Yet it was the gigantic crowds, the largest most of them had ever seen under one roof, that witnesses repeatedly commented on. The Great Exhibition for which the palace had been constructed had on display more than 14,000 inventions from around the world, ranging from machine tools to pins, and at the opening ceremony, when Queen Victoria appeared to the sound of a gigantic organ accompanied by a massed choir thundering out the Hallelujah Chorus from Handel's *Messiah*, she was followed by an international train of princes, heads of state, and ambassadors. The sight almost overwhelmed William A. Burt, a surveyor from Michigan.

"Dear Companion," he wrote that evening to his wife, Phebe, back in Michigan, "this grand precession consisted of nearly every nation under heven in one harmonious band engaged in one object; the like was never seen before. I cannot give you anything like an inteligent view of it. It exceeds by far anything and everything that I had imagined. There is here everything to be seen in nature & art that this world produces."

Square-shouldered and bearded, Burt had spent the greater part of his life in the emptiest parts of the frontier among the hills and lakes of Michigan and Wisconsin, and nothing had prepared him for this experience. His invention of a solar compass had brought him across the Atlantic. In the Michigan peninsula, enormous iron deposits threw off the land surveyors' magnetic compasses so badly, the needles danced back and forth through as much as sixty degrees. Burt's compass, which used the sun to give true north, was immune to such distortions. Its value had led it to be chosen as one of 560 American exhibits, alongside Cyrus McCormick's mechanical reaper and Samuel Colt's repeating revolver, on display at the Great Exhibition.

The exhibition, however, was dominated by the display of industrial muscle that powered the British Empire. Its star was Sir Joseph Whitworth, whose engines cut and ground out machine parts to a tolerance of one ten-thousandth of an inch. A generation earlier, in 1824, when the standards for the imperial measures were created, that degree of precision had represented the very limit of scientific accuracy; in the intervening years it had become a commercial commonplace. As Hassler had foreseen, accuracy's children were devouring their parents, the finer measurements driving out the rougher. By 1851, Whitworth's measurements, and particularly Whitworth's screws, so dominated metal-cutting technology that they virtually held the Industrial Revolution together.

By that date, too, it was clear that in Britain, unlike the United States, industry had overtaken land as the prime source of wealth creation. To William Burt everything about the two countries was different. "I had a fine view of this ancient country," he told Phebe after his train journey from Liverpool to London, "it is all cultivated like a garden, it is unlike any thing that I have seen in America." Above all there was the social divide, and in his next letter William's pride at finding himself in royal and aristocratic company struggled with the democratic instincts of a midwesterner. "I have had an opportunity to associate with Lords and Noblemen and to be in the presence of Queen Victoria and Prince Albert and other Princes, for whome mutch greate veneration is had in England, *as if the gods were present*," he confided to Phebe, "but I could not feel any such veneration for them, for I saw and felt that they were human beings, but a worm of the dust like myself."

Land gained by the United States via the Guadalupe-Hidalgo Treaty, 1848.

Thomas Jefferson would have approved these sentiments and, perhaps rightly, given the credit to his own scheme of democratic land distribution. But there was much more in common between the hierarchical British Empire and the republican United States than he or Burt might have cared to admit.

*B*Y THE TREATY OF GUADALUPE-HIDALGO, signed in 1848 with its army encamped in Mexico City, the United States forced the Mexican republic to surrender a territory that included present-day California, Nevada, Utah; most of Arizona, New Mexico, and Colorado; and part of Wyoming. Added to the annexation of Texas in 1845, this constituted an area larger than the Louisiana Purchase; an understandably complacent census official pointed out that the addition made the United States "of equal extent with the Roman Empire or that of Alexander, neither of which is said to have exceeded 3,000,000 square miles."

Almost from the moment of independence, an expansionist strand of American thinking had envisoned the new nation that would grow beyond the Ohio as an empire. Thus when Rufus Putnam petitioned Congress in 1783 to allow veterans to settle west of the Ohio, he argued their settlement would be "of lasting consequence to the American empire." In the 1820s Missouri senator Thomas Hart Benton took up the idea, and it was his dream of an empire stretching as far as the Pacific that inspired both John L. O'Sullivan's famous phrase in 1845 about the nation's "manifest destiny to overspread the continent," and the goal of William Gilpin, first governor of Colorado, to establish what he proudly and repeatedly called the "Republican Empire of North America." A railroad and communications center for the United States should be located in the prairies, Gilpin argued in *The Central Gold Region*, published in 1860, where it would become "the cardinal basis for the future empire now erecting itself upon the North American continent."

What underpinned this vision was the effectiveness of the public land survey in making ownership of the conquered land available to U.S. citizens. The discovery of California gold in 1848 drew people to the far side of the continent, but it was the grid that enabled the country to be settled at such phenomenal speed. Had he not been in London, William Burt would have been on the other side of the world, on the Oregon coast, where his friend William Ives was running the Willamette Meridian north from a point near Portland, using a Burt solar compass. Just sixty-six years after Thomas Hutchins's chainmen took their first westward steps on the banks of the Ohio River, the land survey had reached the Pacific, and in 1855 the land office of Oregon registered a claim from John Potter, married man of Linn County in Oregon Territory, for "320 acres of Land, known and designated in the Surveys and Plats of the United States as Part of Sections 22 & 27 T[ownship] 9 S[outh]. R[ange] 2 E[ast]." Most of the country from the Mississippi to the West Coast remained unsurveyed, but the squares now spanned the continent.

The only other imperial power to rival this rate of territorial acquisition during the nineteenth century was Britain. Following the loss of its North American colonies, it had conquered India, occupied Australia and New

Zealand, pushed its Canadian claims to the Pacific, moved into southern Africa and established a string of colonies reaching to the borders of Ethiopia. The need for uniform weights and measures throughout its growing number of possessions impelled the British government to pass the 1824 Imperial Weights and Measures Act.

The two countries shared other characteristics—language, elected government, and a legal framework based on the common law—and in the early nineteenth century they were unique in their greed for territory, their respect for property, and their faith in technology. And both relied on Gunter's chain and the surveyor's map to transfer ownership of conquered land to their own citizens.

Their success was bought at the expense of the original occupants. In New Zealand, British settlers acquired almost 80 percent of the land from the native Maoris, using force in a series of wars up to 1870, and quoting the 1840 Treaty of Waitangi, which granted the government rights of preemption, to justify the campaigns. Under the Winthrop doctrine that they had neither enclosed the land nor kept cattle, Australian aborigines were deemed not to have any rights to the country they had occupied for 40,000 years, and the whole of Australia was treated as an empty land—*terra nullius*—to be occupied as Britain saw fit. In southern Africa, a succession of peoples from the Xhosas near the Cape of Good Hope through the Swazis, Matabele, and Kikuyu as far north as Kenya, saw land they had traditionally occupied turned into property by the surveyors' chains, theodolites, and plats. Although some were allowed to buy this newly defined land, the title to it was held for them by the government. Only whites could own it individually.

Most of the surveys were by metes and bounds rather than squares, but the worst of the chaos that plagued land distribution in the southern states of the United States was avoided. This was largely due to the legal presumption that, as in America before the Revolution, the land ultimately belonged to the Crown. Thus in Canada, Native Americans could cite George III's proclamation guaranteeing them the inviolacy of territory west of the Appalachians, and so claim some protection against the newcomers. More generally, the presumption led to the establishment of a central registry of claims in many parts of the empire, known from its inventor as the Torrens system, and encouraged a slightly more ordered pattern of settlement than the rush for land in America.

Yet in the end, wherever the nineteenth-century surveyors unrolled Gunter's chain and drew maps recognized by English common law as a record of a property claim, they achieved the same result. Only in North Africa, where France transformed Arab land into French estate with the help of *géométristes* using the metric system, was there anything comparable to the spread of the Anglo-American property makers in the nineteenth century. Everywhere it depended on being able to measure and describe the land more precisely than the indigenous peoples could. Where the first inhabitants could meet those demands, their rights might have been respected. Where they could not, they were moved off the land—and what happened to Australian Aboriginals, New Zealand Maoris, and southern African Xhosas bore a marked resemblance to the fate of American Indians.

THE TERRITORY THAT WILLIAM BURT surveyed was originally Ojibwe or Chippewa land. In 1836, shortly before Burt began to run the Michigan Principal Meridian due north to the Upper Peninsula, Henry Schoolcraft, the Indian agent for the Great Lakes region, had persuaded the Ojibwe to sign a treaty transferring the title of their land in northern Michigan and Wisconsin to the United States.

Given the difficulty of running a straight line through drifts of winter snow, and clouds of summer mosquitoes, and depths of permanent swamp, Burt could have been forgiven for wondering, as other surveyors did, why Schoolcraft bothered. One page of Burt's field notes records him entering a

William Burt

marsh at 5.68 chains from his starting point. At 40 chains, where he was supposed to insert a post, his note runs: "No post set nor bearing taken. Water 3 feet deep." When he did emerge onto dry land at 71.20 chains from his starting point, it was to plunge into prickly undergrowth that tore his clothes to shreds.

"Dear Companion," he wrote to Phebe, "I am now more than halfway to Lake Superior from Mackinaw, and about 40 or 50 miles from any Settlement in the midst of a swamp about twelve miles in diameter but expect to get out tomorrow as I can see high Beech and maple Land to the North and a River between [us] and it, but I have just made and launched a Canoe. Nothing that will be news to write, but wish to hear from home. My Coat and Pantiloons are most gone. If you could make me a frock [long coat] like that of Austin's [their son] and a pair of Pantiloons of the strongest kind of Bedticking they would I think stand the Brush."

Since no squatters had arrived in Michigan before them, some surveyors took advantage of the general ignorance to complete imaginary plats and field notes. One whose crimes were discovered pleaded reasonably enough that the deceit was not important "because the land would not be sold in ten centuries." William Burt was not that sort. His conscientiousness can be guessed from a detail. He discovered that in the bitter frost of a Michigan winter, a Gunter's chain would contract by as much as an eighth of an inch. Compared with the errors that arose from the unpredictable swings of the compass and the viciousness of nature, this tiny alteration might have been overlooked; but Burt made it his habit to build a fire each morning so that he could warm up his chain to its full 22 yards.

Driving through the Michigan peninsula today, it takes an effort of imagination to see past the soya fields and groves of black walnut to that raw land beneath; but the north-south, east-west roads that lead seamlessly from neat farms into the neon sprawl of shopping malls and condo developments explain how the transformation came about. Below the roads run the surveyors' lines which squared off the wilderness, and not only made it ready for sale but constructed a shape for county and state government. At three dollars a mile, the survey was one of the great bargains of the century.

However, Burt's meridian and the surveyor's squares obliterated something else, and ironically it is due as much to Henry Schoolcraft as anyone that its value is known today. In a series of monumental studies of Ojibwe culture, Schoolcraft presented a detailed account of a society with an intimate, animistic connection between the human spirit and the natural world. In 1855 Henry Wadsworth Longfellow published his own version of that society's legends in an epic he called *The Song of Hiawatha*.

> *Bright above him shone the heavens,*
> *Level spread the lake before him;*
> *From its bosom leaped the sturgeon,*
> *Sparkling, flashing in the sunshine;*
> *On its margin the great forest*
> *Stood reflected in the water,*
> *Every tree-top had its shadow,*
> *Motionless beneath the water.*

Before the words were published, the survey had marched across Hiawatha's forests, revealing much of it to be commercially valuable white pine that the lumber industry would fell; and in Negaunee on the Upper Peninsula beside one of Hiawatha's lakes Burt had discovered in 1842 the richest lode of iron ore yet found in the United States, which would be mined until the tailings turned the lake orange. On either side of long, arrow-straight roads in northern Michigan and Wisconsin, billboards today alternately advertise Ojibwe casinos and the Gamblers' Anonymous warning, "If you are gambling more than you can afford, you may have a problem." It seems like a kind of revenge on the lines beneath.

THOMAS JEFFERSON HAD ENSURED in 1784 that only the United States could acquire title to the American Indians' land. Explaining how this should be done, he wrote: "There are but two means of acquiring the native title. First, war; for even war may, sometimes, give a just title. Second, contracts or treaty." In the years after the defeated nations of the Western Confederacy were summoned to Fort Greenville in 1795, Indian treaties became almost routine. So certain was the federal government of the eventual outcome that the 1804 Land Act extended the surveyor-general's power "over all the public lands of the United States to which Indian title has been or shall hereafter be extinguished."

Almost every Indian war from Fallen Timbers in 1794 to the final massacre at Wounded Knee in 1890 had its origin in the hunger for land, and nearly every treaty that followed involved the transfer of ownership of land. Because U.S. law was equivocal about the transfer of title where coercion was involved, the policy was to accompany victory with purchase, generally in the form of money, gifts, and the guarantee of an alternative home with well-defined borders. The wording of the treaty would make it clear that the contract was voluntarily entered into, and for reasons other than defeat. Thus the Kaskaskia people were said to be ready to sell their territory beside the Mississippi because they were "reduced by the wars and wants of savage life to a few individuals unable to defend themselves against the neighboring tribes." More land up to the Wabash was ceded by the Delaware because, Jefferson asserted, they desired "to extinguish in their people the spirit of hunting, and to convert superfluous lands into the means of improving what they retain." In the South, the Choctaw decided to sell their land, "being indebted to their merchants beyond what could be discharged by the ordinary proceeds of their huntings." Their Chickasaw neighbors gave up land in Tennessee and Mississippi because, according to the 1832 treaty, they "find themselves oppressed in their present situation; by being made subject to the laws of the States in which they reside."

These reasons were no doubt genuine, and a price was certainly paid, but they camouflaged the reality that the United States was making an offer that could not be refused. The treatment of the Cherokee made that starkly clear. Having ceded more than three million acres in what is now Kentucky and Tennessee by a series of treaties, the nation eventually split into two. In 1817 the western, more traditional half agreed to move west to new lands in Arkansas where they could follow their old ways, only to be forced out ten years later by the arrival of more settlers, and sent farther west to Indian Territory in what is now Oklahoma. The eastern Cherokee chose to become Americanized in order to keep what remained of the homeland that had been guaranteed to them by the earlier treaties. Under their half-Scots chief,

John Ross, they built schools, elected a ruling council, wrote a constitution, and farmed as successfully as any other Americans. By 1828 they were a prosperous community, but their land was wanted by Georgia speculators.

Despite two Supreme Court decisions affirming the Cherokee's right to the land, the federal government under President Andrew Jackson, who first made his name in ferocious campaigns against the Creeks, and Seminole in Florida, forced them to accept yet another treaty, surrendering it in exchange for five million dollars and seven million acres in Oklahoma. When they delayed moving out, they were driven off by the army, and in the course of their pitiful journey to Oklahoma in the winter of 1838, better known as the Trail of Tears, some 4,000 of the 16,000 who started out died from disease, hunger, and bitter cold. The implication was clear: Whether the Indians moved or stayed, accommodated or resisted, the title to their land had to be transferred.

The pattern rarely altered: incursion by small groups of settlers, growing tension, Indian violence, American retaliation, and the intervention of the U.S. Army. In Jefferson's presidency alone, thirty-two treaties extinguished Indian titles to most of the land east of the Mississippi, and the missionary John Heckewelder described in his memoirs the relentless pressure on the Native Americans: "They say that when they had ceded lands to the white people, and boundary lines had been established—'firmly established!'—beyond which no whites were to settle, scarcely was the treaty signed, when intruders again were settling and hunting on their lands!" When the Indians complained, "the government gave them many fair promises, and assured them that men would be sent to remove the intruders by force from the usurped lands. The men indeed came, but with chain and compass in their hands, taking surveys of the tracts of good land, which the intruders, from their knowledge of the country had pointed out to them."

With a mixture of desperation and grim humor, the Seneca chief Red Jacket asked for some breathing space in a famous speech in 1829, much quoted by contemporary newspapers: "Brothers, as soon as the war with Great Britain was over, the United States began to part the Indians' land among themselves. . . . permit me to kneel down and beseech you to let us remain on our own land—have a little patience—the Great Spirit is removing us out of your way very fast; wait yet a little while and we shall all be dead! Then you can get the Indians' land for nothing,—nobody will be here to dispute it with you."

With the passage of the Indian Removal Act in 1830, it became government policy to relocate Indians to land west of the Mississippi. In the North the Shawnee, Wyandot, and Delaware nations were moved west from the Lake Erie region to what is now Nebraska and Kansas. In the South, Creek, Choctaw, Chickasaw, and Seminole columns followed the Cherokee to Arkansas and Oklahoma.

*I*N 1851 on the Pacific coast, Burt's assistant, William Ives, began to run the Willamette Meridian from Portland, Oregon, north to the Canadian border. Eventually, every acre in Oregon and Washington State would be related to it, including the rich coastland along Puget Sound. Some of the land acquired by the United States had belonged to a small tribal group, the Duwamish-Suquamish, who were forced to move. The protest of their chief, Sealth, was imaginatively reconstructed thirty years later by an onlooker, Henry Smith, and in the twentieth century was farther embellished to become a well-known environmental protest: "We do not own the freshness of the air or the sparkle of the water. How can you buy them from us?" Beneath the elaborations, however, Smith's words clearly convey not just the intense bond that Seattle (the American version of the chief's name) felt with the land but also Smith's discomfort at what was entailed in America's "manifest destiny" to fill the continent from coast to coast.

"The Indian's night promises to be dark," he wrote in Sealth's name. "No bright star hovers about the horizon. Sad-voiced winds moan in the distance. Some grim Nemesis of our race is on the red man's trail, and wherever he goes he will still hear the sure approaching footsteps of the fell destroyer and prepare to meet his doom, as does the wounded doe that hears the approaching footsteps of the hunter. A few more moons, a few more winters, and not one of all the mighty hosts that once filled this broad land or that now roam in fragmentary bands through these vast solitudes will remain to weep over the tombs of a people once as powerful and as hopeful as your own."

Chief Joseph

By the second half of the century, Smith's ambivalence was widely shared. It was not only Longfellow's *Song of Hiawatha* that found a readership attracted to its picture of the American Indians' pristine world. Many journalists gave compassionate accounts of the words and demeanor of the people who were losing their land, often using versions given by government officials who were themselves sympathetic. Artists depicted them in noble pose—even the Bureau of Indian Affairs, whose purpose was to move them off their land, covered its walls with romantic portraits of Indian braves and chiefs.

Sometimes the published versions sound embellished—"Tell your people that since the Great Father promised that we should never be removed, we have been moved five times," an anonymous Sioux chief was quoted as saying in 1876. "I think you had better put the Indians on wheels and you can run them about wherever you wish"—but other accounts caught the deep identification of the speaker with the land. When the Willamette Meridian pushed the Nez Perce from their land in Oregon, for example, their leader, Chief Joseph, mounted a long, bloody defense against the U.S. cavalry that ended in 1877 in his capture on Bear Paw Mountain. The U.S. press published his eloquent explanation of why he had fought to hold on to the land where his father had lived and died. "I love that land more than all the rest of the world," said Chief Joseph. "A man who would not love his father's grave is worse than a wild animal."

But the ambivalence did not affect the outcome. Early in Jefferson's presidency, when settlers were beginning to push into Illinois, one of the native inhabitants was presented to him. "I am a Kickapoo," explained Little Doe, "and drink the waters of the Wabash and the Mississippi." But the Third

Principal Meridian ran through there, one of the squares became Township 14 North, Range 5 East, also known as Sugar Creek, Sangamo County, Illinois, where the Pulliam family settled, and the Kickapoo were moved to Oklahoma. Seventy years later, despite Chief Joseph's explanation that it was the land around his father's grave that made him who he was, a Nez Perce rather than a wild animal, the Willamette Meridian moved him out of the Sierras just as effectively as the Third Principal Meridian had shifted Little Doe and sent him and his people down to join the Kickapoo in Oklahoma.

In the 1880s even the Indian Territory in Oklahoma came under pressure from Texan and Kansan settlers. To forestall the threat of bloodshed from squatting by the Boomers, as they were called, the federal government surveyed two million acres of the territory and declared that they were to be given free to the first people to claim them. On April 22, 1889, about 50,000 would-be claimants crowded up behind a line drawn by the army near the railroad track. At noon a gun fired, and they raced forward to stake their claim, on horseback, on wagons, on foot, even on bicycles. Hamilton Wicks, who was on foot, remembered the wild scramble up the hillside with thousands on either side making for the same area. "The race was not over when you reached the particular lot you were content to select for your possession. The contest was still who would drive their stakes in first, who would erect their little tents soonest, and then who would quickest build a little wooden shanty. The situation was so peculiar. . . . It reminded me of playing blind man's buff. One did not know how far to go before stopping; it was hard to tell when it was best to stop and whether to turn to the right hand or the left. Everyone appeared dazed."

Before the gun fired, the land over which the settlers swarmed was Indian Territory that had been surveyed and laid out into sections, halves, quarters, even quarter-quarters. By nightfall it was property belonging to the claimants. The whole history of the public land survey was compressed into those frantic minutes.

SIXTEEN

The Limit of Enclosure

THERE IS NO LONG MEASURE corresponding to our acres," commented Josiah Gregg when he visited northern Mexico in the 1830s. "Husbandmen rate their fields by the amount of wheat necessary to sow them: and thus speak of *fanega* land—a *fanega* being a measure of about two bushels— meaning an extent which two bushels of wheat will suffice to sow." It was a revealing observation because such a unit could only be variable. In other words, the concept of landed property, in the sense of an area exactly measured, mapped, and precisely registered as belonging to one person, had not yet established itself in Mexico. When American law came to be applied to what had been Mexican land, that fact was to acquire undue importance.

Some measures had begun to be standardized as early as 1821 when Mexico achieved its independence, but each province had its own version of the basic unit of length, the *vara*, approximately 33 inches. The difference from one province to another was little more than an inch, but the conceptual gap that it signified became obvious when Stephen Austin received a grant of Texas land from the new Mexican government on which to settle 300 families. Each family was allocated one *ligua* (approximately 7 square miles), and quite deliberately Austin chose to have every parcel of land surveyed under his personal direction, specifying that the *ligua* was 25,000 square *varas* and that the *vara* measured precisely 33 1/3 inches. In American eyes, ownership required exactness.

He was the first of some thirty-five individual contractors, or *empresarios*, to whom the Mexican government made large grants of land on condition that they brought in other families as colonists. Most of the *empresarios* followed Austin's example regarding the definition of the *vara*, and, equally important, in rejecting the idea of surveying the parcels by metes and bounds. Austin had seen the confusion over land boundaries that "still exists in Kentucky, Tennessee, and many other states," and made the critical decision to have his Texas land surveyed "regularly and accurately" rather than "to let each Settler run his lines as he pleased and mark them or not." As the federal land survey had found, rectangles made for simple surveying, and the usual shape of a Texan *ligua* was square or oblong with one side on a riverbank.

Three governments were to have control over the land that Austin and the other American *empresarios* carefully measured out. Neither the Texas nor the U.S. governments ever questioned their title to their property. And it was the attempt of the Mexican government in 1835 to challenge that unquestionable right to their land that drove the Texans to fight for their independence.

When Texas was annexed by the United States in 1845, it retained its public land; but the rectangles that the *empresarios* had drawn earned the state some of the advantages of the federal land survey. By avoiding, for the most part, the litigation-haunted metes-and-bounds system that dogged other southern states, Texas was able to develop a land market that gave the state's economy a vigor its neighbors lacked. Over the next half century, fifty million acres were sold to support education, thirty million given to railroad companies, three million to the builders of the state capital, and the rest distributed as military bounties or allocated to squatters and homesteaders. By 1898, Texas had disposed of all its public land but in the process had transformed much of it into productive capital that helped finance its fledgling oil industry.

The Guadalupe-Hidalgo Treaty had guaranteed landowners their right to property granted by earlier governments, but throughout the region covered by the treaty, differences between Mexican and American understanding of what constituted property caused bitter disputes. The document accompanying a Mexican land grant only noted the main boundary markers, and the plat, or *diseño*, that came with it was not much more than a sketch map. Ownership was created not just by the paperwork but by occupation and communal acceptance. Consequently, in New Mexico and Arizona, the land survey almost came to a halt in its attempts to disentangle fraudulent claims from genuine Spanish and Mexican grants, and the traditional rights of local people to land and water. As late as 1890, the surveyor-general of the territory of New Mexico observed, "Certain title to the land is the foundation of all values. Enterprise in this Territory is greatly retarded because that foundation is so often lacking." Handicapped by the absence of a market in land, the New Mexico economy was slow to develop before the twentieth century, and much of the capital then had to come from outside.

In all the region covered by the Guadalupe-Hidalgo Treaty, it was California that was quickest to establish the essential framework of measurement. On July 17, 1851, U.S. deputy surveyor Leander Ransom struggled to the summit of Mount Diablo, east of Oakland, and dug a hole in what he described as the "haycock shaped" summit, marking the initial point of the first meridian in California.

"From the top of this mountain a beautiful prospect is opened before you," he wrote in his report, and went on to describe the sweeping view from the glitter of the Pacific Ocean in the west, across the tremendous gap of the Golden Gate, around to the Vaca Mountains in the north, and eastward up the Sacramento River and the nearby valleys. "These valleys and the ravines and hills surrounding them are mostly covered with a thick set of wild oats growing from 4 inches to as many feet in height," Ransom noted. "The wild oats afford abundant pasturage to the extensive droves of cattle and horses that are scattered abroad over this magnificent range, and also to herds of elk, antelope and deer that abound here."

By the 1860s, when the survey based on the Mount Diablo Meridian covered much of northern California, the gold fields were largely exhausted and the most productive acres were the drained, fenced farms and orchards of the Sacramento and San Joaquin Valleys. Of the 813 grants of land made by the Mexican government, almost one in three had been rejected, but the advantages to the remainder of becoming American property owners with the ability to borrow against the value of the land and increase its fertility could be seen in the appearance of their farms. Earth once given over to cattle had been improved to the point where cereals like rice and wheat could be grown, and in more favored places there were orchards of plums and apples or rows of vegetables. The survey gave farmers secure title to their land, but it also allowed government to function efficiently. As early as 1860, California's surveyor-general was urging the U.S. surveyors to define the border with Nevada as quickly as possible. "There are many settlers in the valley along the border who have never paid taxes," he wrote. "They are undoubtedly in California. The amount of taxable property is considerable amounting to several millions." All that was needed to make the money California's was a line drawn in the sand.

"Surveying in California is a different operation in many respects from what it is in the other States of the Union," Ransom reported, and explained that gold fever had inflated wages so much that even a humble chainman expected to be paid up to $100 a month, compared to $15 in the prairie states.

Yet the link between Ransom's straight lines and the farms in California's Great Central Valley cannot be missed. The valley runs roughly northwest to southeast, but the citrus orchards, lettuce fields, avocado groves, vineyards, and asparagus beds are all aligned strictly by the cardinal points of the compass, north to south and east to west. The yellow dirt tracks that form their boundaries appear at regular one-mile intervals following the surveyor's lines in a neat rectangular plaid.

The same fight with geography occurs in Los Angeles, which should be tilted west of north to align itself with the coast and with the Santa Monica Mountains. To an outsider, the contest is bewildering. Sometimes geography wins, and streets run parallel to the shore or a ridge; sometimes the victor is the survey. Wilshire Boulevard, for example, starts off looking like a geography street as it runs northwest from the city center, but then a hidden force suddenly pulls it due west, and the whole area on either side swings over, desperately trying to align itself with the invisible meridian and baseline. Over most of Long Beach, the San Fernando Valley and central Los Angeles, the meridian wins out, but between Beverly Hills and the ocean, geography comes into its own, and on the fault lines between the two systems, strangers can lose themselves for weeks.

A TRAVELER ON A FLIGHT from Los Angeles to Kansas City, touching down in Phoenix, Arizona, can catch a glimpse of what makes the U.S. land survey one of the most astonishing man-made constructs on earth. In all the flat land east of Los Angeles, the same struggle can be seen to fit squares of housing into valleys and canyons of every shape but square. The artificial pattern nearly disappears in the desert and dusty red sierras, but high up in the mountains it emerges again in patches of cultivated bottomland where the edges of rectangular fields are aligned with the cardinal points of the compass. All at once, looking down through the clear air, you can imagine the surveyor's straight line, drawn west to east along the base of one of those fields, running invisibly over the bare rocks and dry earth, then coming into sight again blacktopped, graded, rollered, and ruled plumb down the center of a street in Phoenix. And when the aircraft takes off heading east again, the line is still there, stretching ahead as a street, along a rank of shopping malls, faster than a Boeing's shadow, the edge of an industrial park, the limit of an executive housing development, a desert track, straight as a die, suddenly ending in red cliffs and obliteration, until 10 or 100 miles farther east, wherever people have settled, it is reborn as a section road or the boundary of a trailer park.

"It was then in a kind of way that I really began to know what the ground looked like," wrote Gertrude Stein in 1937, taking her first flight over the United States after years in Paris where the straight-edged images of Picasso and Mondrian had educated her eye; "quarter sections make a picture and going over America like that made any one know why the post-cubist painting was what it was."

The aircraft eats up in less than three hours the 1,300 miles from California to Missouri that the public land survey teams crawled over for more than

Aerial view of South Dakota today.

half a century. The showpiece of their efforts lies in the Great Plains, where the checkerboard of squares permeates the landscape, the economy, the very outlook of those who live there. To surveyors this is the area controlled by the Fourth, Fifth, and Sixth Principal Meridians. Over land that is never quite flat, but rarely rises to bluffs more than a few hundred feet high, and mostly billows gently like a gigantic sheet in the breeze, surveyors like W. J. Neely found the ideal raw material for their art. Whereas poor William Burt sometimes struggled to make 40 links in a day through the Upper Peninsula, Neely regularly covered 480 chains, or 6 miles, a day as he surveyed "rolling, first-rate prairie, destitute of timber" in South Dakota. Drawn through Minnesota, the Dakotas, Iowa, Nebraska, and Kansas, the surveyors' right angles became the dominant feature of the land.

South from Fargo, North Dakota, the section roads cut each other at exactly one-mile intervals. The ground is so level that a pickup truck traveling a parallel road a mile away seems to float, red metal above gold stubble on black earth. In his book *Take the High Road*, written in 1939, the pilot Wolfgang Langeweische wrote approvingly that it was "just what a pilot wants a country to be—graph paper. You can head the airplane down a section line and check your compass. But you hardly need a compass. You simply draw your course on the map and see what angle it makes. Then you cross the sections at the same angle."

West of the Ohio River, most state boundaries are defined either by rivers or by meridians and parallels, but in determining the shape of the Great Plains territories, Congress was so influenced by the land survey that, wherever politics and geography permitted, it made the straight line king. It carved out two states, Wyoming and Colorado, to a perfect box shape, each being four geographic degrees high and seven degrees wide; it awarded three degrees of longitude each to Kansas, Nebraska, and South and North Dakota, then stacked them up like drawers in a filing cabinet between the thirty-seventh parallel and the forty-ninth. Alongside them to the west, it gave four degrees of longitude each to the drier states of Colorado, Wyoming, and Montana, and piled them up from the thirty-seventh to the forty-ninth with the same geometric regularity. Only Montana's western border with Idaho betrays a wanton crookedness.

What neither the map nor the view from an airplane can reveal is the tension between these artificial shapes and the environment. From the Côteau des Prairies, a long escarpment marking the edge of glaciated plains in South Dakota, there is a view of squared-off prairie—fields, farms, windbreaks, section lines—stretching to the northern and western horizons and all obeying the survey; the sheer expanse of it is as moving and terrifying as an army on parade. Yet even on the stillest day, another power makes itself felt. The dry grasses rustle with it, an insistent pressure of the air, no heavier than breathing, that comes from far off and passes with irresistible momentum from the North to the warm South. The very gentleness of it is sobering. Out there a breeze would shake you on your feet, a storm would knock you flat. It carries the heft of a continent.

The violence of prairie weather was something appalling to settlers. At first it seemed that life could hardly survive the seasonal extremes of temperature, with 40 degrees or more below freezing in winter, and weeks when the summer heat never dropped below 100 degrees. When the first arrivals did begin to farm, nothing had prepared them for the spectacular tornadoes and blizzards that burst out of a clear sky and wheeled in giant vortices through the open land. Only after they had been there for some years did they realize that the real threat lay not in these furious eruptions but in the slower, imperceptible pattern of drought and rainfall. According to Lorin Blodget's groundbreaking study *Climatology*, published in 1857, the 100-degree meridian running through the Dakotas, Nebraska, Kansas, and Oklahoma marked the critical boundary—on a rainfall map the land to the west of it was labeled ominously "the Desert Plains."

In 1862 the Union Pacific Railroad was authorized by Congress to start building westward from "the hundredth meridian of longitude from Greenwich" near Omaha, Nebraska. It was financed by government bonds and by grants of public land up to 20 miles deep on alternate sides of the track—the total came to almost thirteen million acres. In that same year of 1862, President Abraham Lincoln fulfilled a campaign promise by signing into law the Homestead Act, which enabled anyone to acquire 160 acres of surveyed land simply by settling there and improving part of it for five years—that is, by building a cabin and plowing the soil. A fifteen-dollar filing fee then made the homesteader a property owner. The history of the prairie states was shaped by the conflict between those two events and the climate.

Land speculation had never gone away, despite a disastrous slump in 1837, but the railroads brought it back in spectacular fashion. The boast of William Ogden, president of the Union Pacific, was calculated to encourage others to follow his example, but it was also probably true: "I purchased in 1845 property in Chicago for $15,000 which twenty years thereafter was worth ten millions of dollars. In 1844, I purchased for $8000 what eight years thereafter sold for three millions of dollars and these cases could be extended almost indefinitely."

What these profits concealed was the fate of earlier owners. In Sugar Creek, 180 miles southwest of Chicago, few of the squatters who had moved onto the Kickapoo's land in the 1820s succeeded in buying their holdings after Angus Langham had drawn his squares. Most packed their wagons and drove away. Yet the settlers who had patented and paid for claims to a 40-

mile or a quarter-section hardly did better. At one point, surly, stump-legged Robert Pulliam, the definitive pioneer, owned 560 acres—his original 480 and ten years later another 80 on the other side of the creek, where he planned a dam for a new water mill. It should have been the classic transition from land to business, but he had paid for it all with borrowed money, and when his lenders called in the loan, his property evaporated.

It reappeared in the hands of people like the shrewd and careful farmer Philemon Stoute Jr., who in 1846 inherited 350 acres from his father, and in 1881 owned 2,300 acres in Sangamon County. His success came from hiring hands to do extra work and investing his profits in quarter-sections, which he rented out to tenant farmers. Most of his hired help and his tenants were supplied by the families of settlers who had lost their farms following the 1837 slump. Jefferson's democratic dream was already beginning to look like Hamilton's capitalist hierarchy when the railroad companies and the Homestead Act reignited the old ideal.

East of Chicago, the railroad tracks ran between cities, but to the west they were being extended into empty land. The companies needed people to fill it, to produce cattle and crops for transportation, to push up the value of railroad acres, and, with the right sweeteners, to buy them in preference to the government's free 160-acre quarter-sections. "You can lay track through the Garden of Eden," James J. Hill, founder of the Great Northern Railroad, pointed out. "But why bother if the only inhabitants are Adam and Eve?"

As an inducement, the companies offered a dream based on the "forty." A railroad forty could be had for $100 cash, or $25 down and the balance at 6 percent interest over three years. It made more sense than homesteading. The land was close to the track. The railroad provided a flat-packed, prefabricated cabin. It was easy to ship produce out and bring goods in. But the forty was just a grubstake. Like prudent bank managers, the railroad advertisements urged settlers to start with the minimum and "build up gradually." There could hardly be any doubt that they would succeed. "Why emigrate to Kansas?" ran an advertisement in *Western Trail* published by the Rock Island Railroad. "Because it is the garden spot of the world. Because it will grow anything that any other country will grow, and with less work. Because it rains here more than in any other place, and at just the right time." The warnings in *Climatology* could be ignored. As the soil was cultivated, it released moisture—"The rain follows the plow" was the phrase used.

"Four thousand and Four Bushels of Corn from One Hundred Acres," the Chicago and Northwestern Railroad company boasted of its land in South Dakota. "Alfalfa is the best mortgage-lifter ever known," wrote a Great Northern publicist. "It is better than a bank account for it never fails or goes into the hands of the receiver. It is weather-proof, for cold does not injure it and heat makes it grow all the better. . . . For filling a milk can, it is equal to a handy pump. Cattle love it, hogs fatten upon it, and a hungry horse wants nothing else." In an inspired piece of advertising, the Northern Pacific claimed that not only would a Montana farm make its owner rich, but the state's dry climate was so healthy that not a single case of illness had been recorded there in the previous twelve months—except for indigestion caused by overeating. The Chicago and Northwestern came back with a poster in big red letters proclaiming the superior fertility of Dakota: "30 Millions of Acres of the Most Productive Grain Lands in the World. You Need a Farm." Entering into the spirit, the Canadian Pacific shouted louder still about Alberta's rich black soil: "Our lands should yield you annually 100% of the purchase price; and besides they should increase in value at the rate of 20% a year for at least 40 years."

At a time when Hollywood was nothing more than arid scrubland, the railroad companies were demonstrating the power of a dream. From all over Europe, people responded to the promise that prairie land, north or south of the border, would not only guarantee independence but prove to be a license to mint money. In Kansas, Carl B. Schmidt, an agent of the Atcheson, Topeka and Santa Fé Railroad, brought over no fewer than 60,000 German settlers himself, while agents of the rival Kansas Pacific arranged for the steamer passage and railroad journeys of thousands more from Scandinavia, Britain, and Russia. But by far the greatest number came from within the United States—the restless, landless, ambitious residents of Illinois, Iowa, Wisconsin, and other already occupied states to the east. By the end of the nineteenth century the railroad companies had sold about 120 million acres, comfortably exceeding the 80 million acres of free homestead land that were claimed in the same period.

The reality the settlers encountered was not the rural idyll suggested by the advertisements. In the dry or semiarid lands beyond the corn belt, it quickly became apparent that no one could live off a railroad forty or indeed a homesteader's quarter. It was too dry to grow crops unless there was a creek or a borehole for irrigation, and running cattle required vast acreages to provide sufficient grazing—5 to 8 acres per animal was the minimum. A succession of acts—the Desert Land Act, the Timber Culture Act, and others—allowed larger areas to be homesteaded so long as the settler irrigated the soil or grew trees that were thought to generate moisture themselves. They could not alter the bleak reality that the model of the small farm on which the survey was founded could not work in such conditions.

*I*N 1902 Grace Fairchild from Wisconsin followed her footloose husband, Shiloh, to a homestead in the dry land of the Dakotas. The economics of the farm depended on being able to put cattle out on the unclaimed open range, and to cut hay from bottomland that belonged to no one except the government. But as more and more homesteaders came in, fencing off the range, there was less free government land available. "We settlers only had 160 acres in the early days," Grace wrote in a memoir, "and that is not enough land to support many critters or make a living raising cash crops. We had to increase our holding or get out."

Shiloh, whose dream was to breed horses, was as impractical and ambitious as Robert Pulliam back in Illinois, but he had a wife who would have delighted the souls of both Jefferson and Hamilton. The former would have appreciated the way she started a school for her nine children, read books, increased corn yields with new strains of seed, bought a good bull to improve the herd, and showed sufficient independence of mind to keep her improvident husband farming rather than horse-ranching. Hamilton would have approved the way Grace earned extra money keeping pigs and chickens, selling butter, taking in guests—parties of up to sixteen would-be settlers herded by land agents—as well as bottling, barreling, and pickling the produce from her garden, on top of caring for her large family. Like everyone who reads her story, both would have been astonished at her stamina.

With droughts lasting long enough to kill the deep-rooted alfalfa, bliz-

Grace and Shiloh Fairchild

zards coming as late as May, and plagues of grasshoppers, which usually arrived "when the drought had us against the wall," the chance of a wipeout hung over every Dakota farmer. The harsh truth of homesteading in dry country was that survival did not depend just on stamina or good luck but on others' failure. Abandoned claims offered free pasture and hay to the farmers who remained. When the land was sold to pay back-taxes, anyone with spare cash or credit, like Grace Fairchild, could acquire it cheaply. After forty years, her homestead had become a ranch of 1,440 acres, and she had put most of her children through college on the proceeds.

Grace consciously sought out the wider horizons that education offered, but even for her a prairie farm threatened to become a prison. In the early days, she noted, "most of our neighbors lived 90 miles east of us in Fort Pierre, or twelve miles south of us on the Indian reservation." Even when the land filled up, her nearest neighbor was a mile away, and her best friend, Mrs. Kurzman, who helped deliver several of her babies, had a claim 4 miles distant. "In no civilized country have the cultivators of the soil adapted their home life so badly to the conditions of nature as have the people of our great Northwestern prairies," wrote E. V. Smalley in the *Atlantic Monthly* in 1893. "Each family must live mainly by itself, and life, shut up in the little wooden farmhouses, cannot well be very cheerful. . . . An alarming amount of insanity occurs in the new prairie states among farmers and their wives."

What Smalley blamed were the squares. These encouraged settlers to build their houses near the center of their holdings, so that no field was too far away. As a result, on two neighboring 160-acre claims, there would often be half a mile between farmhouses as the crow flew. But the isolation was intensified by the failure of the survey to allocate land for roads. Inevitably these developed along the section lines, and so in practical terms, for one farming family to reach another required a journey down the track to the section line, along the section line, and up the neighbor's track, a distance closer to a mile. Smalley wanted all the owners of farms in four quarter-sections to "agree to remove their homes to the center of the tract" and thus create a village with their land redistributed outward from the center. Such an arrangement would, he argued, attract people "of such a sociable, neighborly disposition as would open the way to harmonious living."

In seventeenth-century Massachusetts, even in eighteenth-century Marietta, that system of in-lots and out-lots radiating out from a town or village center would have seemed obvious. The pattern had changed while the survey was still releasing the forestland east of the Mississippi. With a square of woodland to clear, most settlers started at the center. A popular sequence of prints published in the 1850s by Currier & Ives expressed their ambitions. The first picture showed a wooden cabin surrounded by the forest; in the next the cabin has grown to a little cottage standing in a clearing; by the third it was a farmhouse with a yard in front and behind it rectangular fields lined by woodland; and the last, titled *The Land Is Tamed*, portrayed a two-story mansion surrounded by farm buildings, a squared-off garden, and straight-edged fields stretching away to the horizon. No other building was in sight.

The contrast with Canada was particularly striking. In Quebec and Ontario, French and British land distribution tended to produce farms in the shape of long thin oblongs, similar to the long lots in Louisiana. The narrow end usually fronted onto a river or road, encouraging settlers to build houses close to this line of communication with their farmland stretching out behind. Thus, instead of the typical American farmhouse in the middle of a square section of land with the nearest neighbor a mile or more away, the

A Currier & Ives print celebrating the settlement of the wilderness.

Canadian farmer usually had seven or eight neighbors within half a mile on either side. Even in the prairie provinces of Alberta, Saskatchewan, and western Manitoba, where the land was surveyed and squared off into 640-acre sections subdivided into halves and quarters, the Canadian survey had allowed a strip of 1½ chains, or 99 feet, for roads between the townships. The existence of this shared highway tended to draw houses toward it, shortening the distance between families. South of the forty-ninth parallel, the U.S. prairie farmer endured a uniquely solitary existence where progress depended on an individual's ability to impose his or her will upon the land.

The cost was not only human. Square farms carved out of the country regardless of drainage slopes were vulnerable to erosion, and it was from the dry plains, plowed and planted to exhaustion when prices were high during the First World War, that the soil blew away in the great dust storms of the 1930s that darkened the daytime sky 1,500 miles away. That catastrophe represented the most visible sign of the reckless exploitation of the land's re-sources. In the East, great forests had been burned to clear the soil for crops; the buffalo, which had roamed as far as Georgia in herds 10,000 strong, were on the edge of extinction by 1875; and the passenger pigeons, which once flew in flocks so great they took hours to pass and darkened the sun, had vanished altogether.

When nineteenth-century conservationists like John Muir and John Burroughs began to argue for the social and spiritual benefits of using the land more gently, they faced a particularly American obstacle. While most Western nations had land laws restricting individual property rights in favor of social needs, the United States had the opposite. Combining the legal con-cept of "fee simple" with the Fifth Amendment—"nor shall private property be taken for public use, without just compensation"—the law protected the property owner from almost all government interference so long as taxes were paid. To a degree unknown in the rest of the world, Americans were monarchs of their property, entitled to do almost what they pleased—until it infringed on the rights of other property owners. "There is as yet no ethic dealing with man's relation to land and the animals and plants which grow upon it," wrote Aldo Leopold in *A Sand County Almanac* in 1949. "Land, like Odysseus' slave-girls, is still property. The land relation is still strictly eco-nomic, entailing privileges but not obligations."

Yet the sheer beauty of a place like the Yosemite Valley made the conser-vationists' argument for them. When it became the first U.S. National Park in 1890, it was a symbol that a different way of thinking about the land was emerging. There the wilderness was more than about-to-be property; it con-tained "natural curiosities or wonders," as the park's definition put it, worth keeping for their own sake. In 1906 Teddy Roosevelt made the idea a gigantic reality when he set aside almost 200 million acres of the remaining public domain for forest and national parks. For ranchers, and lumber and mineral companies, this new category of land was and remains an anomaly. Having been surveyed and measured, it ought to have become property and been used for farming, timber felling, or mineral extraction. Yet for the first time since Thomas Hutchins crossed the Ohio River, the pattern had been broken, and the tension of that underlying conflict still surrounds the public lands, and the Bureau of Land Management which is responsible for them.

$\mathscr{T}$HE MAGNITUDE OF the greatest land-measurement project in history is mind-boggling," wrote the geographer Hildegard B. Johnson in 1976. "Never was so much land surveyed in so short a time under the same standardized methods. . . . one marvels at the determination with which these men threw and retraced their lines. Still their role as civilian heroes of the frontier is largely ignored in the history of the frontier."

There were mistakes and frauds. Disputes over the line of the California-Nevada border continued until the 1980s. The redwood area of northwest California was imaginatively surveyed by John Benson and a syndicate of accomplices working in the bars of San Francisco with the help of maps bought from the Coast Survey, and others did the same in Colorado and Utah. Nevertheless, the survey in the forty-eight lower states was virtually complete by the 1930s. Most of Alaska had still to be measured—and parts are still unsurveyed—but the great majority of the surveyors' work was concerned with checking the accuracy of original surveys or replacing wooden posts and other markers that had rotted, been burned, or otherwise disappeared.

Since 1785 the landmass of the United States has grown to 2.3 billion acres, and of that total, 1.8 billion acres spread across thirty-two states have been at one time in the public domain. More than one billion acres have been transferred to individual ownership, with seven million remaining in state and federal government hands. In economic terms alone, it has represented the greatest orderly transfer of public resources to the private sector in history.

Somewhere west of the Côteau des Prairies, on a line running down through the dry-land prairie, there is a boundary in that history. This is the open range once occupied by the Sioux and the buffalo, and then by the drover-cattlemen whose semiferal longhorns grazed its short grass before being rounded up for the stockyards. In the 1870s an estimated eleven million cattle roamed free on what was effectively common land. By then the railroads were bringing in the range's new owners, but lack of timber prevented them from putting up rails to mark out their property, though some grew fences of Osage orange trees.

In 1873, Henry Rose, who had a farm 60 miles west of Chicago, invented barbed wire, and immediately it became possible for the range to be enclosed along the lines shown on the surveyors' plats. In 1880, 50,000 miles of fence were in place, and more than 200,000 tons of wire were being sold each year. Within another five years, sixteen million acres had been fenced in, and the homesteaders' wagons loaded with wire and posts leapfrogged past existing claims out into the short-grass prairies. In 1890, the superintendent of the census reported that "up to and including 1880 the country had a frontier of settlement, but at present the unsettled area has been so broken into by isolated bodies of settlement that there can hardly be said to be a frontier line."

Turner recognized this as the end of frontier life, but an older history also halted there. As the farmers planted their barbed-wire quadrangles across the range, they came into ferocious conflict with the range cowboys. Such a battle could have only one outcome. The entire edifice of the law—and with it every sheriff and U.S. marshal who ever became a white-hatted movie hero—was on the side of property; in other words, of those whose land was measured and entered on a plat. Accepting the inevitable, cattlemen carved out ranches from the open range, fenced them in, and started to haggle with the Bureau of Land Management over grazing rights on the public land that remained.

The nomadic range cowboys were the last in a line of unpropertied people to claim the land—and to be defeated by the law. By the 1890s they were on their way to join the Mexican pueblo farmers, the Native Americans, the South African Xhosas, the Scottish Highlanders, and Elizabeth I's sturdy rogues and vagabonds among the dispossessed. The long march of enclosures that had begun in England in the 1530s had reached its culmination 350 years later in the dry-land prairies of America.

SEVENTEEN

Four Against Ten

FOUR HAD BEEN THE KEY TO IT ALL. That was the number that linked the sides of a township to the old medieval units of the rod and the acre. It was a number the mind could immediately recognize, and a quantity, as Jefferson understood when insisting on square containers, that could be easily visualized and simply tested. To the land dwellers who relied on them, the 4-based measures felt organic, almost instinctive, a sound, practical, hands-on, symmetrical way of measuring; and with this went a way of thinking that valued those qualities. In the West the idea of being "a four-square guy," or of being offered "a square deal," came to epitomize everything that was desirably solid and reliable; and by the same way of thinking a square represented all that the unpropertied, sinuous world of jazz despised.

"Foure graines of barley make a finger," was the rule of Queen Elizabeth I, "foure fingers a hande; foure handes a foote." And by a freak of history, those ancient quantities had left their mark on the character of the most modern society on Earth. Edmund Gunter's chain, the first U.S. standard of measurement, carried the old English system of land measuring, the 8-furlong mile, the 4,840-square-yard acre, and the 4-based way of thinking, into every corner of the land and planted it deeply in the American psyche.

Confronted by the spread of English measures, Ferdinand Hassler had standardized the other weights and measures that went with them: the 16-ounce pound, the 8-pint gallon, and the 4-peck bushel. As each square township out on the frontier filled with people and enough squares within a territory petitioned for statehood, an exact, precision-machined replica of the federal government's 4-square weights and measures was sent to the new state capital.

Among the huge variety of state laws and constitutions, this national system of weights and measures created one single market for American industry's goods. It permitted manufacturers to standardize and mass-produce everything from agricultural machinery to typewriters and sewing machines. For those who wished to acquire wealth, industry became the new frontier. In just twenty years, between 1879 and 1899, the value of industrial products almost tripled, to thirteen billion dollars a year. The fortunes acquired by Andrew Carnegie from steel and John D. Rockefeller from oil proved beyond doubt that the day had passed when land was the prime source of productive wealth.

As it had been at the very start of the industrial age, the ability to measure precisely was a source of power. In the 1760s James Watt had struggled to get tolerances of a hundredth of an inch, but before the end of the century Henry Maudsley, the British designer of the screw-cutting engine lathe, refined the benchmark to one-thousandth of an inch, and in the 1850s Joseph Whitworth's routine production was measurable in ten-thousandths of an inch. Metal, power-driven lathes working at this tolerance created the power looms and locomotives that drove forward the second stage of the Industrial Revolution. By the 1880s Pratt and Whitney were supplying a market for commercial machines capable of checking precision gauges to one hundred-thousandth of an inch.

That degree of accuracy was good enough for the metal-bashing end of industry, but the late-nineteenth-century technologies of electricity, telegraphy, chemicals, and pharmaceuticals demanded even higher standards. To be used effectively, their quantities had to be assessed with extraordinary fineness.

"Although the electrical is the latest developed branch of engineering, it is the most exact in its measurements," a technical journal boasted in 1903, "currents from as small as 1/10,000,000 ampere and less, up to 2,000 or 3,000 amperes being easily and accurately measured. . . . in electrical work, [these are] powers large enough to operate a train of some hundreds of tons weight, or powers so small that no ordinary electrical device could measure them." The precision this implied across a huge range of power output was breathtaking, but it was not just the exactness of the measurement that would have shaken a chain-wielding surveyor to the steel-capped toes of his boots. The unit used, the ampere, or amp, contained a secret—it was metric.

*T*HAT THE METRIC SYSTEM had survived at all was astonishing. For the first thirty years of its existence, it had seemed perpetually on the point of being abandoned. The revolt against it had begun in France almost as soon as the law of 18 Germinal Year III, or April 7, 1795, had made it the only legal measure that could be used. "The centuries-old dream of the masses of only one just measure has come true!" the Committee of Public Instruction proudly announced. "The Revolution has given the people the meter." Whether they wanted it was another matter.

Despite the publication of numerous tables giving equivalents between the old and the metric measures, the country was utterly unprepared for the new system. Customers could not understand the metric units, and traders immediately took advantage of the change to increase profit margins. In the great agricultural market of Les Halles in Paris, grain merchants measured sacks of oats by the small bushel but charged them at the price of a large hectoliter. The same trick was played in bakery and grocery shops, where two sets of scales and measures were used, and goods would be sold in old measures under metric names, or vice versa—whichever paid best.

Where there was no fraud, there was confusion, particularly when it came to decimals. In the fashion shops, where lengths of cotton and linen had been sold by the *aune* or ell, it ought not to have been difficult to substitute the meter, which was only a little longer; but the system contained a snakepit of lesser numbers. Everyone knew what half an *aune* or a quarter of an *aune* looked like, and they could halve or quarter the meter in the same way. Given time, they could even estimate how long half a quarter of a meter should be, but no one, not even the ink-stained bookkeeper upstairs, knew that it was the same as 0.125 of a meter, or, still more bafflingly, 12.5 centimeters.

An alarmed government circular blamed the unpopularity of the system on criminals. "The general public suffers from dishonest practices," it claimed in 1797, "and so loses faith in a system that seeks above all to benefit it." But as the *cahiers de doléances* had made clear, what the public wanted was the old system made uniform, one *toise*, one *aune*, one *quintal* throughout the nation. Instead they had been given what Witold Kula, the Polish science historian, called "a strange new-fangled measure allegedly relating to the very land people walked but in a manner that no one understood." The Greek and Latin names, all the kilos and centis that the scientists loved, simply sounded so foreign that French patriots refused to say them, asking instead for the old *livres* and *aunes*.

The foreignness of the metric system went deeper than names. It took uniformity to a degree that no layperson could immediately comprehend. The traditional measures had variety because they related to different activities. Cloth was measured by the ell or the *aune* because it was natural to hold it and stretch out the arm to full length. A journey was measured by the yard or the *toise* because the road was walked. Land was measured by the acre or the *arpent* because that represented work. The metric system forced people to separate the measure from the activity altogether and deal with an abstract unit that, as Kula observed, "would be equally applicable to textiles, wooden planks, field strips and even to the road to Paris." What underlay the popular dislike of the metric system was a very modern anxiety, the sense of alienation from the natural world.

In 1799 the metric system received endorsement from a French-led international commission made up of client states like Switzerland (represented by Hassler's friend Johan Tralles), Batavia, and Lombardy. Following their approval, permanent standards for the meter and kilogram were made from Joseph Dombey's platinum based on the measurements of the meridian by Delambre and Méchain. These standards were deposited in the Archives of the Republic, and copies were made for members of the commission. When Napoleon seized power and crowned himself emperor in 1804, he threw all his power and prestige behind the meter. An edict issued in the following year announced, "It is most definitely his unalterable wish to maintain the new system of weights and measures in its entirety and to accelerate its extension throughout the Empire." But neither international prestige nor imperial decrees could persuade the French to love the meter.

The emperor did not have much time for decimals himself. In his army the official daily ration for cavalry horses remained a quarter of a *boisseau*, the artillery measured the caliber of its guns in *pouces* and *lignes*, and when his military engineers crossed the Berezina River into Russia, they estimated its width at forty *toises*. By then, Napoleon had begun to retreat in the face of France's massive sullen resistance. In February 1812 his much-feared minister of the interior, Jean-Pierre de Montalivet, informed the prefects in charge of France's eighty-three departments that the emperor had come to recognize that "metric units are not suited to practical daily needs. The exclusive employment of the decimal system may suit book-keepers but is by no means well adapted to the daily dealings of the common people who have much difficulty in understanding and applying decimal divisions."

The solution was to introduce what was called the *système usuelle*, combining metric and organic units. The *toise* came back, but measuring 2 meters, as well as 6 *pieds*. The *livre* reappeared, representing both 500 grams and 16 ounces, while the new *pied* was either 33 centimeters or 12 *pouces*.

When Napoleon was overthrown and the Bourbon monarchy restored in the shape of Louis XVIII, the metric system, brought in by the hated Jacobins who had guillotined the last king of France, might well have been abolished. It had almost no friends, yet somehow it survived. The explanation lies in a snobbish aside made by the vicomte de Chateaubriand in his *Mémoires*: "Whenever you meet a fellow who instead of talking of *arpents*, *toises* and *pieds*, refers to hectares, meters and centimeters, rest assured, the man is a prefect." And in case any of his readers had forgotten what a prefect was, the vicomte helpfully defined him as "a person characterized by petty tyranny, a bureaucrat concerned with conscription"—and, it should be pointed out, with responsibility for weights and measures.

Among the briefing papers put in front of Louis XVIII after his coronation in 1815 was one from his new minister of the interior, to whom all prefects reported. "Sire!" it began. "The uniformity of weights and measures has long been desired in France, and your royal predecessors sought to establish it. I presume therefore that your Royal Highness will wish to uphold an institution that accords so excellently with his great ideas of public utility and whose effective spread is in any case by now very much advanced."

Whatever the new king's private opinion, he was a realist. He might have claimed to be an absolute monarch, but he ruled through the vast, centralized government machine that had grown out of the Revolution and the Napoleonic Code. It was the machine that wanted the metric system. As Napoleon himself had been forced to admit, the simplicity of calculating in decimals suited bookkeepers—and prefects, and bureaucrats, and government officials of every kind. Swallowing his reservations, Louis agreed to keep the metric system alongside the customary measures. "The important lesson France has taught the world," remarked a cynical friend of Kula, "is the effectiveness of centralized administration."

It was not wholly a coincidence that the unpopular metric system spread across Europe in concert with the growth of government bureaucracy. Each of the client states that first received the metric system—Lombardy, Switzerland, and Batavia, which was to become part of Belgium—reacted in the same way as France, accepting it officially and as a symbol of modernism and democracy, while rejecting it at almost every other level. But like France, they discovered that the system was a one-way street. Introducing it created chaos; trying to go back to the customary measures produced still more. The indecision lasted for a couple of generations, and not until the 1850s did their governments find the courage again to make the meter the only official measure.

It was Prussia that best learned the French lesson about centralized administration and thus set the model for metrication. Napoleon had lost faith in the system by the time he crossed the Rhine in 1806, but in 1868 the autocratic Prussian prime minister, Otto von Bismarck, set about modernizing the most traditionally hidebound state in Germany. In what was called "a revolution from above," he reformed its administration, its public education, and its social security—and replaced the old *Zoll* (inch), *Pfund* (pound), and *Morgen* (half-acre) with the centimeter, kilogram, and hectare.

Once Prussia, the first truly modern state, succeeded in unifying Germany beneath its banner, the meter and modernism soon spread into Austria and Hungary. Beyond German borders a freshly united Italy caught the spirit and embraced the metric system as a symbol of its new nationalism. Scandinavia had adopted it earlier, in 1863, and Spain, which had introduced it for the second time without much enthusiasm, extended it to Cuba and Spanish America, where it took hold among newly independent countries enthusiastically signaling their progressive credentials. In 1870 the delegates from twenty-four nations met in Paris to agree on international standards for the meter, and even the French, who in 1840 had had it reimposed as their country's only legal system, became—rather doubtfully—a little proud of it.

*T*HUS, AGAINST ALL THE ODDS, the deeply unpopular units based on Borda's triangulation had survived and prospered. But the system had other allies than bureaucrats. While the chain had been spreading traditional measurements through the American West, the physicists mapping out the vacant properties of heat, light, and electricity had decided to measure them metrically. They needed simplicity and exactness, and in their work, pounds and feet were impractical.

The challenge came from the dimensions of the world that science uncovered in the century after 1785. It grew until it encompassed everything from the unimaginably small to the inconceivably large, from molecules to galaxies. To describe these newfound extremes created a need for precision beyond anything that had gone before. In the United States, Hassler found little support for his efforts to push the standards of accuracy to extraordinary limits, but in Europe the École Polytechnique in Paris, and universities like those at Berlin, Glasgow, Turin, Uppsala, and Cambridge, as well as scientific bodies like the Royal Society and the Académie des Sciences, created an environment in which precise measurement was recognized as fundamental to scientific experimentation and discovery.

Its importance prompted the great Glasgow physicist William Thomson, later Lord Kelvin, to make his famous comment in a lecture in 1891: "When you can measure what you are speaking about, and express it in numbers, you know something about it; but when you cannot measure it, when you cannot express it in numbers, your knowledge is of a meagre and unsatisfactory kind: it may be the beginning of knowledge, but you have scarcely in your thoughts, advanced the state of science." Thus, just as the surveyors had once used measurement to create property from the wilderness, so in the nineteenth century the physicists used it to create science from the natural world.

But first, one fundamental need had to be met. To describe how electrical energy is turned into heat, light, magnetic energy, or mechanical power required a new system of uniform measurement. The problem was considered in 1861 by a committee of the British Association for the Advancement of Science, including Kelvin himself as well as two other giants of nineteenth-century physics, James Clerk Maxwell and James Prescott Joule. Any measurement of electrical energy had to be a synthesis of its constituent elements of charge, current, voltage, and resistance. The committee decided that each element should be measured separately, using the fundamental principles of length, mass (as weight was now more accurately described), and time. In what was to be a critical intellectual decision, the members of the British committee chose as the units of length and mass not the inches and ounces they used in everyday life but the meter and gram, later amended to the centimeter and gram. Although time continued to be measured in old-fashioned seconds still reckoned as 1/86,400th part of a day, electrical measurement became metric.

The choice was a matter of convenience. A common system of measurement was needed because physicists from at least four countries had helped uncover the properties of electricity and magnetism. And the ease of converting small quantities to large simply by moving a dot made decimals ideal for measuring the great range of quantities found in the use of electrical power.

The decision of the British Association for the Advancement of Science crowned the metric system as the favorite child of science. Other areas of physics concerned with light, heat, and radiation needed measurement, and all were linked by Maxwell's equations on electromagnetic waves. A series of international scientific congresses hammered out agreement on the nature and names of the new units during the late nineteenth and early twentieth centuries. As well as distance, mass, and time, four other basic elements were recognized—temperature, substance, light intensity, and electric current—for which the units were named, respectively, the kelvin, mole, candela, and ampère. From these seven elements, composite units were derived that could be applied to every other measurable force—power, radiation, electric charge, magnetic strength, and others—many of them named for the scientist or mathematician who first described what was being measured. Units like the watt and the volt, ampere and ohm, curie and hertz, siemens and becquerel, were the equivalent of physics' hall of fame. And they were all metric.

It was the decimal division and the integration of length and mass that proved to be the system's strength. Its origin, as one ten-millionth of the length of a quadrant meridian, had become unimportant. The true distance from equator to pole, now reckoned to be 10,001,965.7 meters, is almost 2 kilometers longer than first estimated in 1794, and more than 1,200 meters beyond Delambre's best computation. Even in the nineteenth century, the 1799 bar of platinum in the Archives of the French Republic, from which all metric lengths were derived, was understood to be as much as 1/150th of an inch shorter than it should be. Nevertheless, where the physicists led, others followed, and for similar reasons.

In 1867 surveyors and mappers from around the world met at a conference in Berlin of the International Geodetic Association and recommended that a new international meter based on the 1799 meter be adopted as the standard for all their work. Governments could ignore science, but maps, which involved frontiers and national territory, were a different matter. A preliminary series of meetings in 1870 was interrupted by war, but the representatives of thirty nations met again in Paris in 1872, examined the meter and kilogram made from Dombey's platinum and kept in the Archives, and decided they should be the basis for a new set of internationally accepted standards. Three years later, eighteen countries signed the Treaty of the Meter, which brought into being the International Bureau of Weights and Measures that still supervises metric standards, and created for it an international enclave, a kind of Vatican of the meter, at Sèvres outside Paris.

The new international standard was introduced in 1889 in the form of a bar cast in a platinum-iridium alloy, but as close in length to the 1799 meter as possible. Instead of being sawed to a length of one meter, with the consequent risk of contracting to a shorter length, it was 102 centimeters long, with two scratches etched into it at exactly one meter's distance. Since it had an error margin of as much as one twelve-thousandth of an inch, it was outdated almost as soon as it was made, and before the end of the century physicists were unofficially defining the length in the more precise terms of wavebands of light. As the world continued to expand relentlessly into subatomic quantum particles and stellar regions beyond the Milky Way, the need for a still more exact definition grew urgent.

It came in 1960, with a new specification based on the wavelength of light. Careful measurement of the 1889 meter, itself based on the 1799 bar, showed its length to be equivalent to 1,650,763.73 wavelengths of radiation emitted by the orange krypton-86 atom. This became the new definition of a meter. The accuracy was subatomic, a margin of error much less than one-millionth of an inch; but for the first time there was no physical standard against which a length could be checked, merely a scientific experiment.

The change was critical. Once released from physical constraints, the meter has become defined with a growing fineness that now far exceeds the imagination of the nonscientist. It is presently equal to the length of the path traveled by light in a vacuum during the time interval of 1/299,792,458th of a second, where a second is equivalent to 9,129,631,770 oscillations of the 133 Cesium atom. This may not be quite as straightforward as a yard once was, but it has some advantage in exactness. Using an iodine-stabilized, helium-neon laser built according to international guidelines, one can check the meter's accuracy with an error margin of plus or minus 0.000,000,000,000,02 millimeters. Ever more precise and less comprehensible definitions of its length will certainly continue to evolve.

Only the kilogram remains a physical reality, an iridium-platinum cylinder made, like the meter, in 1889 and kept in a locked vault in the Sèvres headquarters of the International Bureau of Weights and Measures. It usually remains out of light and sight, protected against deterioration by three airtight bell jars. Although it looks remarkably like the copper example that Joseph Dombey carried with him on the *Soon*, it is no longer a unit of weight but of mass.

For most practical purposes, weight and mass are virtually the same, but as scientists are quick to point out, if you were on the moon your mass would not alter, but your weight would be only one-sixth of the amount on earth. This is because weight measures the force of gravity on mass. Even on earth, a person increases in weight at the pole, where gravity is stronger, and floats more lightly at the equator, where it is weaker. Thus, by the sort of logic that can blow a fuse in the nonscientific mind, the weight of the kilogram is actually determined in newtons, a unit of force (one kilogram is roughly 9.8 newtons), while its mass is simply "the mass of a particular cylinder . . . which . . . is preserved in the care of the International Bureau of Weights and Measures in a vault at Sèvres, France."

Eventually this, too, will be consigned to history, to be replaced by a definition related to the electrical force needed to lift a kilogram, or the number of atoms in a silicon crystal of that mass. The substitute is needed because, for all its protection, the mass of the Sèvres kilogram is getting larger as a film of atmospheric particles coats it, adding what is effectively the mass of a grain of pepper every decade.

At the eleventh General Conference on Weights and Measures held in 1960, all the metric measures, old and new, were brought together into what is called, in English, the International System of Weights and Measures. Its official title, in deference to the nationality of Jean-Charles Borda, with whom it originated, is Le Système Internationale de Poids et Mesures, or SI for short.

That year marked the point at which the globalization of measures became a reality. The government of every industrialized country in the world, including the United States, signed the Treaty of the Meter. All their weights and measures were defined in relation to the meter and the kilogram. Most had gone farther and adopted the SI as their official system.

In 1785 James Madison had written prophetically that a scientific, decimalized system "might lead to universal standards in these matters among nations. Next to the inconvenience of speaking different languages, is that of using different and arbitrary weights and measures." Much of that inconvenience had been removed, and the massive expansion in international trade following the Second World War was undoubtedly accelerated by the single language of measures that by 1960 had come to be spoken around the world. In two major areas, however, it still had to be translated into a local patois. In those countries that Winston Churchill dubbed "the English-speaking democracies" and General Charles de Gaulle simply called "the Anglo-Saxons," the old traditional measures were still firmly in place.

EIGHTEEN

Metric Triumphant

BETWEEN AUSTRALIA, CANADA, New Zealand, South Africa, the United Kingdom, and the United States, there is indeed a bond that arises from language and form of government, but they share a third and perhaps more fundamental tie. One of the profoundest influences on their legal and democratic systems has been that concept of private landed property whose physical reality was carved out with Gunter's chain. To protect what they claimed as theirs against overly greedy rulers, landowners had struggled to establish a system of individual rights and a form of government in which their interests were represented. Thus for all six nations, property and democracy became entwined at the same point, the enclosures in Tudor England, and they seemed inseparable from the system of organic, 4-based weights and measures on which ownership and exchange depended.

By contrast, the meter had emerged from the same intellectual ferment that enunciated the rights of man as an idea, something innately human and existing independently of property or any other tangible quality. Around the world the metric system had always been introduced from the top rather than in response to popular demand. Each country had its own motives—the Soviet Union went metric in 1918 as part of a program of cutting ties with its old imperial Russian past; in 1947 a newly independent India announced its intention to switch in order to underline its break from Britain; Japan changed in 1958 in accordance with an overall plan to expand its foreign trade—but there was always one constant: the existence of strong central government with a clear desire for modernity.

As colonial powers, nineteenth-century France and Belgium made the metric system part of what they deemed their civilizing role in African possessions such as Algeria and the Congo, while former British colonies like Ghana, Nigeria, and Kenya adopted it in the twentieth century as a symbol of their newfound independence. In 1872 the liberal emperor Pedro II introduced it as part of his program to modernize Brazil's economy. Mexico had the system imposed three times in a decade—twice by Benito Juárez and in between by the French puppet emperor Maximilien—each time in the name of progress. When Mao Zedong and the Communist Party decreed in 1959 that the People's Republic of China would go metric at the start of their ambitious but disastrous modernizing plan, the Great Leap Forward, the system circled the globe.

For the English-speaking holdouts, the question of adopting the metric system consequently carried a wider significance than the simple exchange of one set of measures for another, and for a century every attempt by governments to move toward introduction of the meter had been frustrated.

Despite its resistance to Jefferson's decimals, the United States ought to have been one of the first metric countries. As early as 1866, just nine years after establishing the American Customary System of Weights and Measures as the federal system, Congress passed the Metric Act allowing metric measurements to be legally used alongside it. In 1875 the United States became one of the original signatories to the Treaty of the Meter, and an increasingly influential scientific and technological community pushed the government toward adopting the system. In 1893 Thomas C. Mendenhall, the politically powerful boss of the Coast and Geodetic Survey and thus, following Hassler's precedent, responsible for weights and measures, persuaded the Treasury that the yard should no longer be defined by the 1857 standard but in relation to the decimally divided, light-wave-determined, scientifically based meter. At that point, the intellectual battle should have been over—the American Customary System no longer had any independent basis for its standards.

When politicians appeared more inclined to listen to the argument that a change would confuse customers, particularly the poor and the less educated, the scientists voiced their opinions with growing impatience. "If the introduction of the metric system is to be accomplished in America," a correspondent wrote sharply to *Science* magazine in 1894, "we must act in the light of experience already acquired in Europe which is far more valuable than any amount of theorizing about the effect upon poorer classes who have not yet tried it."

Any growth in popular sympathy for the metric system was effectively halted in the first half of the twentieth century by the two World Wars. It was in the aftermath of victory, in 1945, that Winston Churchill coined his phrase "the English-speaking democracies," but he might equally well have named them the acres-and-ounces democracies. Having rescued the world from the threat of totalitarian dictatorship, their populations saw less need than ever to succumb to metrication.

In 1959 the governments of the six agreed to harmonize their measures in relation to the metric system. (A pound was declared to be equivalent to 0.453,592,37 kilograms, and one inch to 2.54 centimeters; the impossibility of making an exact equivalence led to a 3-millimeter difference between the international mile, derived from a metric base, and the statute mile of the United States, which is derived from the foot used in the public land survey.) But while the change bound the systems of the six closer together, it exposed a fundamental weakness.

The United States already defined its measures in relation to the meter, but in Britain and the Commonwealth nations, the ultimate standards of their weights and measures were the metal bar and cylinder constructed in 1845 to replace those melted when the Houses of Parliament burned down. By the 1950s, the latest scientific methods revealed that not only were there minute inaccuracies in their construction, but exposure to the atmosphere was causing a steady, microscopic deterioration in their condition. Either new standards would have to be made or a new definition established. In 1963 Britain bowed to the ravages of time and legislated to define its yard in relation to the meter and the indestructible, invariable wavelength of orange-red krypton 86. When Australia, Canada, South Africa, and New Zealand followed suit, every major country in the world shared one scientific definition of its weights and measures.

Thereafter the chasm grew wider, as the governments of each of the six holdouts, urged on by their scientific, business, and industrial communities, began to adjust for formal entry into the metric system, while the majority of their populations opposed it.

In 1965 the British government announced a ten-year plan for conversion, and until 1979 a Metrication Board announced the different areas of business, industry, and the administrative structure that had gone metric—glassmakers, schools, government contracts, filling stations, the London Metal Exchange. The alcohol in whiskey was expressed as a percentage rather than as proof; the gill in which it was measured followed the pottle and the mutchkin into history. Even Gunter's chain, the very measure that made private property possible, faded from view, accompanied by a bodyguard of rods, poles, and perches.

The conversion was voluntary, but in 1973 Britain joined the European Common Market, whose rules were designed to make the metric system compulsory. The period of conversion was elastic, in some cases lasting more than twenty years, but sooner or later the change had to be made. By the time it was wound up in 1979, the Metrication Board reported that except for retail food stores, home improvement sales, and road signs, metric measures were either replacing traditional units or being used alongside them. Almost as an aside, its final report noted that just 31 percent of the population supported the change.

A similar sequence of events occurred in Australia, Canada, New Zealand, and South Africa. The process was voluntary up to the point where customers could make their wishes felt. Thus prepackaged food like breakfast cereals and frozen peas was sold in metric units early on, but fresh food like potatoes and fruit continued to be weighed in pounds and ounces. Wholesale quantities in the timber trade were specified metrically, but individual planks were ordered and sold in feet and inches. To clear up the confusion, retail trade associations and chambers of commerce began to urge governments to move into this area too, and to pass legislation making the old units illegal. Australia, New Zealand, and South Africa had largely completed the process by 1980, and Canada was judged to be 60 percent complete.

At first the United States marched in step, beginning with a Department of Commerce report to Congress in 1971 optimistically titled *A Metric America, a Decision Whose Time Has Come*, which called for a program to make the country predominantly metric within ten years. Acting with a speed unparalleled in earlier attempts to reform weights and measures, Congress passed the 1975 Metric Conversion Act, setting up a board to "coordinate and plan the increasing use of the metric system in the United States." In an interesting example of political ambivalence, the administration of Ronald Reagan closed down the board but approved the 1988 Omnibus Trade and Competitiveness Act designating the metric system as the "preferred system of weights and measures for United States trade and commerce." For good measure the act also required federal agencies to use the metric system in procurement and other business by a "date certain and, to the extent economically feasible, by the end of fiscal year 1992."

Under the act's impetus, large sections of industry that depended on government contracts went metric, and defense procurement in particular abandoned traditional measures. At the other end of the price scale, soft drinks were sold in 2-liter bottles, and spirits that had once appeared in fifths and quarts came in 750-milliliter and liter bottles. Medicines were measured in grams rather than grains. In the year 2000, the New York Stock Exchange began replacing the old eighths and sixteenths fractions of a dollar in which stocks were priced, and—logically enough—made use of Jefferson's decimals to mark their price in dollars and cents. New sectors like the computer industry quoted quantities of information bytes in metric notation, going from kilo through mega and giga, eventually reaching exa, zetta, and the so-far-ultimate yotta, or 1,000,000,000,000,000,000,000,000 bytes.

According to the writer Robert Sabbag, one part of the economy had already gone decisively metric a decade earlier. "Whatever Congress decides," he wrote in his account of the drug trade, *Snowblind*, "the truth is this: The United States of America effectively converted to the metric system in or around 1965—by 1970 there was not a college sophomore worth his government grant who did not know how much a gram of hash weighed. . . . off the top of his head he could go from grams to ounces, and he could tell you how many ounces he got to the kilo. . . . And so today, everyone over the age of twelve knows there are 28.3 grams to the ounce, and 35.2 ounces or 2.2 pounds to the kilogram."

Throughout the English-speaking democracies the momentum for change seemed irresistible. Running two systems of measurements was inconvenient for government, expensive for industry, liable to create error in science and advanced technology. All of these fields had an interest in taking the last step in metric conversion and eliminating traditional measures; since those measures had already been defined in terms of metric measurement, there was no logical reason for keeping them.

Yet even in countries that had adopted the metric system for more than a century, where children grew up educated in it and thinking in terms of a single unit of length, weight, and capacity, a residue of organic measures remained stubbornly in existence. In German markets they sold, and still sell, meat by the *Pfund*, or pound, meaning 500 grams or 17.6 ounces. In French markets the same unit appears again, this time as the *livre*, while the Danes order in *punds*, the Dutch in *ponds*, and the Swiss in *Pfunds* or *livres* depending on which canton they live in. Heavier quantities such as coal or lumber have kept the hundredweight, or *Zentner* (one-twentieth of a ton), alive in Germany, where it is used in place of 50 kilograms, and the metric ton becomes 20 *Zentners*.

The inch, whose death sentence was signed in Scandinavia in 1863, is resurrected every day by plumbers, who routinely specify the diameter of pipes and faucets in *tumme* (literally thumbs) if they are Swedes, or *tomme* if Norwegian. Even in Europe's industrialized heartland, Germany, the thumb, or the *Zoll*, is what craftsmen use for small measures. Informally, a Swedish farmer will measure his land not by the hectare but by the *tunnland* (roughly 1.2 acres), his Austrian counterpart by the *Joch* (1.4 acres), a German by the *Morgen* (0.6 acres), and an Italian by the *acro* (1.0 acre).

The feature common to all these measures is that they fit everyday activity better than the equivalent metric units, and in personal, hands-on transactions, customers prefer the most suitable, whole-number units. But their survival is due to more than convenience. People insist on using them in direct exchange because it is an unconscious sign of belonging to the community, of knowing the local dialect, as opposed to a stranger who would use the metric term. Here the metric system's strength, its invariable, universal uniformity, becomes its weakness. Like all forms of globalization, it saps the sense of local autonomy.

Popular resistance among the six holdout nations naturally went farther, and in the 1980s it forced the governments of Britain, Canada, and the United States to abandon plans to impose the system compulsorily. In states like Iowa, Alabama, Missouri, and Illinois, where speed limits had been designated in kilometers in the 1980s, the signs were removed in the 1990s. State building codes went from metric back to feet and inches.

In Canada, where metrication had been promoted aggressively with massive publicity in the form of pamphlets, leaflets, and films, and a concerted program covering industry, government, and even retail stores, the public reaction halted the campaign in its tracks in 1983, leaving a remarkable confusion of units. Specifications for construction work on government plans were metric, but for private buildings in feet and inches. Weather temperatures were referred to in Celsius, oven temperatures in Fahrenheit, snow in centimeters, wind in miles per hour. The *Washington Post* reported in 2000 that in Canada's largest supermarket chain, "the produce clerks speak in pounds; the fish counter measures by grams; and the meat department maintains a studied bilingualism—frozen turkeys in metric, fresh in imperial. Loose mushrooms are priced in pounds, but the scale available to weigh them measures only in grams."

A similar confusion reigned in Britain, with television weather forecasts given in metric and Fahrenheit, and doctors recording babies' weights in kilograms but telling their mothers in pounds and ounces. Road signs remained in miles per hour, but gas was sold in liters. In 2000, when Steven Thoburn, a market trader in Sunderland in northern England, refused to quote prices for bananas in kilograms, the city authority took him to court; but elsewhere other would-be "metric martyrs" were allowed to ignore European Union rules, and a reliable estimate suggested that altogether about

30,000 shopkeepers had neither replaced nor converted scales that weighed only in ounces and pounds.

"Their slipping back into the old ways served some hunger of the soul," a commentator in the Australian magazine *Quadrant* suggested of that country's metric refusers. "It might also have been a healthy instinctive resistance to the bossiness of all revolutionaries; that rigid insistence on conformity with the new dispensation."

The refusers found an unexpected ally in the new American-dominated computer industry, whose hardware, such as disks, monitors, and printers, was measured in traditional units. Software for printers and for complex CAD (computer-aided design) applications also used inches and fractions of inches, which created unexpected problems in metric mode. American architects, for example, who use tolerances of one-quarter, one-eighth, or one-sixteenth of an inch in their work, found the software for technical drawing ideal for their requirements, but metric architects complained that the software insisted on approximating their more precise millimeters to the nearest fractional equivalent, and so threw off their calculations. Even the computer's basic information code, based on binary arithmetic and thus having to be counted in 2s, defies decimalization, so that there are 1,024 bytes rather than 1,000 in each kilobyte of information. A recent attempt to replace the megabyte with a new decimalized unit, the mebibyte, has been rejected by traditionalists, who refer to it as a "maybe byte."

In the 1980s the American space industry took an equally robust view in the face of a sustained effort by the National Aeronautics and Space Administration at the start of the international space station program to convert its suppliers to the metric system. "When the time came to issue production contracts," recalled Burt Edelson, a former associate administrator of NASA, "the contractors raised such a hue and cry over the costs and difficulties of conversion that the initiative was dropped. The international partners were unhappy, but their concerns were shunted aside."

The possible consequences of that decision were not to be appreciated for more than a decade, but elsewhere the growing confusion about which units to use produced more immediate results. In 1983 an Air Canada plane in flight from Edmonton to Montreal ran out of fuel and had to glide to an emergency landing in Manitoba after the ground crew had mistakenly measured the fuel payload in pounds instead of kilograms. In Britain there was a tragic outcome involving a newborn baby with heart trouble. The weight of Benjamin Adams was recorded as 7 pounds and 1 ounce, and the doctor treating him was advised to prescribe 10 micrograms of the heart drug Digoxin per kilogram of body weight. Translating imperial weight to metric, the doctor miscalculated by a factor of 10, and ordered a nurse to give Benjamin 320 micrograms of the drug instead of 32. Thirteen hours later the baby died from the overdose. Even though a life was lost, the circumstances—a tired doctor, an urgent case, a snap decision—could be explained. The next mistake was so meticulously considered, so amazingly expensive, so incredible, it astounded the world.

At precisely 5:01 A.M. on September 23, 1999, NASA's press office issued its first release on the critical phase of its fourth expedition to Mars. The spacecraft, launched more than nine months and 416 million miles earlier, was about to enter orbit around the planet, and the release announced that "orbit insertion has begun!" The exclamation point was a tribute to the intricate engineering and computer work involved, as well as to the approximately $250-million price tag for the probe, including $125 million for the craft and its instruments.

The spacecraft was intended to circle the planet, taking observations of its atmosphere from about 65 miles altitude—hence the name Mars Climate Orbiter. As the engines fired to slow it down, the craft disappeared behind Mars, and ground engineers prepared for its return at 5:26 Pacific Coast Time. It did not reappear on schedule, and the stream of news from the press office grew more somber in tone as its absence grew longer. Then, late in the day came the first admission that the craft was lost, and an investigation began into the reasons. It took barely ten days to track down the root cause: "failure to use metric units in the coding of ground software file, 'Small Forces,'" to quote the deeply embarrassed NASA release.

The clash of two systems had produced a childish mistake. The new team taking over responsibility for the Orbiter at the manufacturer, Lockheed Martin Astronautics, was accustomed to dealing in American Customary units rather than metric as specified by the Jet Propulsion Laboratory, which was responsible for managing the Mars missions. In writing software for the Orbiter's thruster engine, used to navigate the craft, the Lockheed engineers specified its force in pounds instead of newtons.* Thus each time the thruster fired to stabilize the craft, it was emitting barely one-quarter the force that the instruments on earth recorded. Orbiter was already more than 100 miles lower in trajectory than it should have been when it went into orbit, and the probability is that it burned up or crashed into the surface of Mars.

Asked for his reaction to the news, Noel Hinners, in charge of flight systems at Lockheed, replied with a comment echoed by most of the scientific community: "The reaction is disbelief. It can't be something that simple that could cause this to happen."

The fact that it had happened provided stark evidence of the state of the United States's measurements. Two hundred years after George Washington pointed out the need for the country to have a single uniform system of weights and measures, it has got two. "Twenty years ago," Hinners said in exasperation, "we went through this whole hassle of, 'Should the U.S. go metric?' I wish we had."

In Britain that point was finally passed on January 1, 2001, when the last relaxation of European Union rules on weights and measures ran out and it became illegal to sell goods by the pound or inch or any other customary measure. The rules allowed market traders to display the old units until 2010, although with their metric equivalent shown more prominently, but after that date complete uniformity will be compulsory in Britain.

If it were a simple question of logic, the United States would also be planning to change. Logically, the American Customary System is now a museum piece, the last remaining relic of an organic, 4-based system that has been preserved across the centuries by a freak of history. It evolved from hands-on, purposeful activity, and every argument that Jefferson put forward about the greater arithmetical simplicity of calculating in 10s still

---

*The equation to translate from American Customary to metric units is 1 pound-force = 0.45359237 kilograms × 9.80665 meters per second squared = 4.448 newtons, approximately.

holds. From the moment that the majority of people started to spend most of their lives indoors, and to earn most of their living not from their hands but from mental activity, the advantages of doubling and halving began to dwindle away.

Logic, however, has never played more than a small part in the history of weights and measures. The rest has been about the distribution of power. In its rawest guise, greater accuracy has given empires the power to explore new areas and to exploit them at the expense of the less accurate. But measurement is also about the power of society to allow a just exchange of goods and cash, and at its most fundamental level it has, like language, the power to express a personal value between the individual and the material world. This is what makes the choice faced by the United States so dramatic.

"The American style has never been to impose radical changes after state commissions decide on their superiority," observed Edward Tenner, a visiting researcher in the history of science and technology at Princeton University. "Americans even hate seeing dual mile and kilometer road signs. The metric system has been a casualty of its identification with political authority."

On the other hand, the ability to measure more accurately has always delivered an economic and informational advantage to those who possess it. And no state that has started down the metric road has ever been able to draw back.

To add to the tension, the European Union has fixed the year 2010 as the deadline after which it will accept only metrically measured goods. Like everything to do with measurement, the EU's decision has a little to do with logic—a single system is more efficient than two—and much to do with control, profits, and power. All that has happened in the last two centuries points to an inescapable decision. Eventually, the U.S. government will have to make a choice between economics and its past, between the wishes of business and those of its citizens.

The Witness Tree

THE CALIFORNIA OFFICE of the Bureau of Land Management is a white, government-designed box on the outskirts of the city of Sacramento. It serves as the depository for the survey plats and notes on which all modern real-estate ownership in the state is based. It attracts a wide range of citizens, from suburban homeowners tangling with neighbors over garden boundaries, to ranchers pressing for grazing rights and gold miners establishing claims for a promising outcrop of mountain near Death Valley. If ever there was a place where measurement and ownership come together, it is here.

"When you see how easy it is to use the land survey," declares Lance Bishop, chief of the BLM's geographic services in California, "you have to admire Thomas Jefferson's foresight in choosing a grid. Every parcel of land has an identity. As an example, I've just bought a 5-acre parcel, and I can go back to the original records and see the shape of the original property, where it was first platted, where the original markers were set, and all subsequent records. It's very clear; there's no ambiguity about what you own."

It is here, too, that a preview of the upcoming battle over measurements can be found. At every level except property, surveying has gone metric. The process began in 1817 when the Coast Survey adopted metric measurement, and now the National Geodetic Survey uses it throughout the United States. The Geological Survey has employed it since 1879; the world's ruling body, La

Fédération Internationale Géomètre, since 1865; and today even the global positioning systems and transits used by every amateur surveyor give measurements in metric units. The figures, however, are translated not just into feet and inches but into chains and links.

"Chain attachment is very strong," admits Bishop. "No one knows what it means any more, but no one wants to give it up. We're pushing it—I've tried platting two sections in metric units, but you would not believe the resistance—it is terrific. And you can understand why."

As early as 1876, the Franklin Institute of Philadelphia listed as a major obstacle to the introduction of the metric system the fact that "the measurements of every plot of ground in the United States have been made in acres, feet, and inches, and are publicly recorded with the titles to the land according to the record system peculiar to this country." Other objectors made the same point, that the legal problems of redefining property would be endless. "Nothing can be more perfect than the land measures of the United States in its square mile sections," Coleman Sellers, a Cincinnati engineer, declared in 1895, "all of which would be thrown into endless confusion by the proposed change of the mode of measuring, without the slightest gain to any human being by the operation."

It need not have been like this. If Jefferson's decimal system had been accepted in one of its forms, the decimal pound and foot would have been at the center of a system of weights and measures that, backed by American industrial muscle and the universal preference for them as units, might have become the world's favorite system. The historic turn was missed, and, left without competition as the only scientifically based, decimalized measurement, the French meter has taken its present place as the simplest, most accurate means of measuring everything between the dimensions of a quark and a black hole.

Now the old battle between 4 and 10 has returned, and the arguments from the original eighteenth-century debate have become modern. Yet this time much more is at stake. Two centuries have so embedded the traditional measures in the land and in the outlook of its inhabitants that they could not be eradicated without the destruction of something irrecoverable. The United States is a democracy built upon a concept of property that owes everything to measurement. If, in the long run, Joseph Dombey and his copper standards do finally arrive, the values that are lost will have come from the heart of the nation's history.

$\mathcal{S}$IX THOUSAND FEET UP in the Sierras, Ed Patton is in search of history, a corner post hammered into the ground in 1873 by Deputy Surveyor William Minto to mark the point where the southeast corner of Section 36, Township 22 South, Range 36 East, Mount Diablo Meridian met its neighboring section. Bearded and ponytailed, Ed is a public land surveyor, employed by the Bureau of Land Management and, as he proudly insists, in direct line of succession from Thomas Hutchins and the men who crossed the Ohio River in 1785. He is conducting a resurvey of the area. It is needed because 130 years ago the surveyor running the line north toward Minto's corner belonged to the Benson syndicate, which produced hundreds of false surveys, and the point of intersection marked on his plat owed more to guesswork than bootwork.

This is not commercial land. It is steep mountain slope made up of dry, stony earth that slides beneath your feet. All that grows up here is sagebrush and stands of piñon pine whose branches make dark-green clumps against the red-gray rubble. Far down in the valley, however, there is good grazing, and it was the need to delineate where the graziers' property ended that initially sent Minto and his team across these harsh, magical mountains. The ground is littered with obsidian chips, some shaped into blades and arrowheads by earlier inhabitants who had no need of marks to know that this was their place; but Patton's mind is on his predecessors.

"They ran a straight line through all this, straight forties and eighties [chains], over mountains, canyons, places so steep you need someone to support you while you sight through the transit," he says admiringly. "And that survey is the one we all go back to. When you find one of their original corners, it is like a handshake with the past."

Using Minto's notes and the scars of blazed trees, Ed navigates along a ridge, slithers into a dip, and then, halfway up a rock-covered slope, comes upon a giant, yellow-trunked piñon reaching out from the past. Carved into

Minto's witness tree.

the living wood and half obscured by a ragged rim of bark is a series of runic incisions:

T ⴻ IIS R ⴻ VI E
S ⴻ VI

Patton runs his finger gently around the mark and translates the runes: Township 22 South, Range 36 East, Section 36. "That's Minto's blaze," he says softly. "When he carved it, this tree would have been no more than twelve inches across. Look at it now. And it'll be here another century."

This is a witness tree, a record of a human claim to part of nature. In 1942 Robert Frost took the name for the title of a collection of poems about the attachment of the American people to their land. At the inauguration of President John F. Kennedy in 1961, Frost himself read a poem from the collection, *The Gift Outright,* whose theme is the way that a country molds its inhabitants. It was their hunger to possess this particular land, Frost suggests, that enabled the land to perform its magic and transform those who hungered into Americans.

This land was ours before we were the land's.
She was our land more than a hundred years
Before we were her people. She was ours
In Massachusetts, in Virginia,
But we were England's, still colonials,
Possessing what we still were unpossessed by,
Possessed by what we now no more possessed.
Something we were withholding made us weak
Until we found out that it was ourselves
We were withholding from our land of living,
And forthwith found salvation in surrender.
Such as we were we gave ourselves outright
(The deed of gift was many deeds of war)
To the land vaguely realizing westward,
But still unstoried, artless, unenhanced,
Such as she was, such as she would become.

—From *A Witness Tree,* 1942

ACKNOWLEDGMENTS

Triangulation, the surveyor's standby, sparked the idea for this book. The first point was a chance reference to Joseph Dombey, the unlucky messenger sent to the United States in 1794 with a copy of the prototype kilogram and meter. The second was the vivid memory of a flight from Los Angeles to New York on a clear winter's day that revealed a checkered land that reached from east of the Rockies to Pennsylvania. The third was a chapter in John Stilgoe's *The Common Landscape of America* describing the effect of the public land survey on the appearance of the countryside. To fill in the ground between those distant points, I relied enormously on the skill, experience, and generosity of many people on both sides of the Atlantic.

In roughly chronological order, I should like to express my gratitude to Malcolm Draper for kindly sharing his vast experience of surveying, demonstrating the use of a theodolite, and at a later stage for reading part of the manuscript; to Steve Booth, editor of *Surveying World*, for his encouragement and supply of vital information; to James Kavanagh of the geomatic faculty at the Royal Institution of Chartered Surveyors, for his friendship and for introducing me to the institution's library; to Richard Johnson, my editor at HarperCollins in the United Kingdom, who not only took the idea seriously but told others about it; to Lyn Cole, who undertook research in France with undaunted pertinacity; to my wife, Marie-Louise, for driving with me through 5,000 miles of American back roads with only one argument, and for opening my eyes to innumerable aspects of the passing landscape that I would otherwise have missed; to Holly Iverson in Fargo, North Dakota, who shared her family history of life in the state; to Catherine Renschler of Adams County, Nebraska, for her personal and historical insights into pioneer life on the prairies; to Professor Tom Schmiedler and Barbara Jarvis in Lawrence, Kansas, for their advice and hospitality; to Lance Bishop, chief of geographic services in the Bureau of Land Management at Sacramento, California, for giving generously of his time and resources; to Ed Patton, heir to the tradition of Thomas Hutchins, who showed me what the public land survey was all about; to Gerard Iannelli of the Metric Program, NIST, for his information about the attempt to introduce metric measurement to the United States; to Lawrence Brooks for advice on quitrents; to Penry Williams for reading the chapter on Tudor history; and above all to George Gibson of Walker & Company, New York, for the kind of dedicated and rewarding editorial colloquy that most authors imagine went out of fashion a generation before they were born.

Among institutional sources of information, I would especially like to thank, for their professional help and the use of their resources, the staff at the London Library, the British

Library, the library of the Royal Institution of Chartered Surveyors, the Campus Martius Museum at Marietta, Ohio, the Michigan Museum of Mining, the State Archives at the Michigan State Museum, the Marquette County Historical Society, the North Dakota Institute for Regional Studies, the Adams County Historical Society in Nebraska, the archives of the Bureau of Land Management, Sacramento, California, the Library of Congress, the library and museum of the National Institute of Standards and Technology in Gaithersburg, Maryland, the New York Public Library, and, for its miraculous speed at accessing information on the Web, Google.

Responsibility for errors of omission and commission, on the other hand, belongs entirely to me.

APPENDIX

General Tables of Units of Measurement

Appendix C of NIST Handbook 44, Specifications, Tolerances, and Other Technical Requirements for Weighing and Measuring Devices

The following tables have been prepared for the benefit of those requiring equivalency units for occasional ready reference.

1. TABLES OF METRIC UNITS OF MEASUREMENT

In the metric system of measurement, designations of multiples and subdivisions of any unit may be arrived at by combining with the name of the unit the prefixes deka, hecto, and kilo, meaning, respectively, 10, 100, and 1,000, and deci, centi, and milli, meaning, respectively, one-tenth, one-hundredth, and one-thousandth.

In certain cases, particularly in scientific usage, it becomes convenient to provide for multiples larger than 1,000 and for subdivisions smaller than one-thousandth. Accordingly, the following prefixes have been introduced and these are now generally recognized:

yotta, (Y)	meaning 1,024	deci, (d),	meaning 10^{-1}
zetta, (Z),	meaning 1,021	centi, (c),	meaning 10^{-2}
exa, (E),	meaning 1,018	milli, (m),	meaning 10^{-3}
peta, (P),	meaning 1,015	micro, (u),	meaning 10^{-6}
tera, (T)	meaning 1,012	nano, (n),	meaning 10^{-9}
giga, (G)	meaning 109	pico, (P),	meaning 10^{-12}
mega, (M),	meaning 106	femto, (f),	meaning 10^{-15}
kilo, (k),	meaning 103	atto, (a),	meaning 10^{-18}
hecto, (h),	meaning 102	zepto,(z),	meaning 10^{-21}
deka, (da),	meaning 101	yocto, (y),	meaning 10^{-24}

Units of Length

10 millimeters (mm)	= 1 centimeter (cm)	
10 centimeters	= 1 decimeter (dm)	= 100 millimeters
10 decimeters	= 1 meter (m)	= 1,000 millimeters

10 meters	= 1 dekameter (dam)	
10 dekameters	= 1 hectometer (hm)	= 100 meters
10 hectometers	= 1 kilometer (km)	= 1,000 meters

Units of Area

100 square millimeters (mm²)	= 1 square centimeter (cm²)	
100 square centimeters	= 1 square decimeter (dm²)	
100 square decimeters	= 1 square meter (m²)	
100 square meters	= 1 square dekameter (dam²)	= 1 are
100 square dekameters	= 1 square hectometer (hm²)	= 1 hectare (ha)
100 square hectometers	= 1 square kilometer (km²)	

Units of Liquid Volume

10 milliliters (mL)	= 1 centiliter (cL)	
10 centiliters	= 1 deciliter (dL)	= 100 milliliters
10 deciliters	= 1 literl (L)	= 1,000 milliliters
10 liters	= 1 dekaliter (daL)	
10 dekaliters	= 1 hectoliter (hL)	– 100 liters
10 hectoliters	= 1 kiloliter (kL)	= 1,000 liters

Units of Volume

1,000 cubic millimeters (mm³)	= 1 cubic centimeter (cm³)
1,000 cubic centimeters	= 1 cubic decimeter (dm³)
	= 1,000,000 cubic millimeters
1,000 cubic decimeters	= 1 cubic meter (m³)
	= 1,000,000 cubic centimeters
	= 1,000,000,000 cubic millimeters

Units of Mass

10 milligrams (mg)	= 1 centigram (cg)	
10 centigrams	= 1 decigram (dg)	= 100 milligrams
10 decigrams	= 1 gram (g)	= 1,000 milligrams
10 grams	= 1 dekagram (dag)	
10 dekagrams	= 1 hectogram (hg)	= 100 grams
10 hectograms	= 1 kilogram (kg)	= 1,000 grams
1,000 kilograms	= 1 megagram (Mg) or 1 metric ton (t)	

2. TABLES OF U.S. UNITS OF MEASUREMENT

Units of Length

12 inches (in)	= 1 foot (ft)	
3 feet	= 1 yard (yd)	
16½ feet	= 1 rod (rd), pole, or perch	
40 rods	= 1 furlong (fur)	= 660 feet
8 furlongs	= 1 U.S. statute mile (mi)	= 5,280 feet
1,852 meters	= 6,076.115 49 feet (approximately)	
	= 1 international nautical mile	

Units of Area

144 square inches (in²)	= 1 square foot (ft²)	
9 square feet	= 1 square yard (yd²)	
	= 1,296 square inches	
272¼ square feet	= 1 square rod (sq rd)	
160 square rods	= 1 acre	= 43,560 square feet
640 acres	= 1 square mile (mi²)	
1 mile square	= 1 section of land	
6 miles square	= 1 township	
	= 36 sections	= 36 square miles

Units of Volume

1,728 cubic inches (in³)	= 1 cubic foot (ft³)
27 cubic feet	= 1 cubic yard (yd³)

Gunter's or Surveyors' Chain Units of Measurement

0.66 foot (ft)	= 1 link (li)	
100 links	= 1 chain (ch)	
	= 4 rods	= 66 feet
80 chains	= 1 U.S. statute mile (mi)	
	= 320 rods	= 5,280 feet

Units of Liquid Volume

4 gills (gi)	= 1 pint (pt)	= 28.875 cubic inches
2 pints	= 1 quart (qt)	= 57.75 cubic inches
4 quarts	= 1 gallon (gal)	= 231 cubic inches
	= 8 pints	= 32 gills

Apothecaries Units of Liquid Volume

60 minims (min or)	= 1 fluid dram (fl dr or *f*)	
	= 0.2256 cubic inch	
8 fluid drams	= 1 fluid ounce (fl oz or *f*)	
	= 1.8047 cubic inches	
16 fluid ounces	= 1 pint (pt or)	
	= 28.875 cubic inches	
	= 128 fluid drams	
2 pints	= 1 quart (qt)	= 57.75 cubic inches
	= 32 fluid ounces	= 256 fluid drams
4 quarts	= 1 gallon (gal)	= 231 cubic inches
	= 128 fluid ounces	= 1,024 fluid drams

Units of Dry Volume

2 pints (pt)	= 1 quart (qt)	= 67.200 6 cubic inches
8 quarts	= 1 peck (pk)	= 537.605 cubic inches
	= pints	
4 pecks	= 1 bushel (bu)	= 2,150.42 cubic inches
	= 32 quarts	

Avoirdupois Units of Mass

The "grain" is the same in avoirdupois, troy, and apothecaries units of mass.

$27\frac{11}{32}$ gains	= 1 dram (dr)
16 drams	= 1 ounce (oz)
	= $437\frac{1}{2}$
16 ounces	= 1 pound (lb)
	= 256 drams
	= 7,000 grains
100 pounds	= 1 hundredweight (cwt)
20 hundredweights	= 1 ton
	= 2,000 pounds

In "gross" or "long" measure, the following values are recognized:

112 pounds	= 1 gross or long hundredweight
20 gross	= 1 gross or long ton
or long hundredweights	= 2,240 pounds

Troy Units of Mass

The "grain" is the same in avoirdupois, troy, and apothecaries units of mass.

24 gains	= 1 pennyweight (dwt)	
20 pennyweights	= 1 ounce troy (oz t)	= 480 grains
12 ounces troy	= 1 pound troy (lb t)	
	= 240 pennyweights	= 5,760 grains

Apothecaries Units of Mass

The "grain" is the same in avoirdupois, troy, and apothecaries units of mass.

20 gains	= 1 scruple (s ap or)	
3 scruples	= 1 dram apothecaries (dr ap or)	
	= 60 grains	
8 drams apothecaries	= 1 ounce apothecaries (oz ap or)	
	=24 scruples	= 480 grains
12 ounces apothecaries	= 1 pound apothecaries (lb ap)	
	= 96 drams apothecaries	
	= 288 scruples	= 5,760 grains

3. NOTES ON BRITISH UNITS OF MEASUREMENT

In Great Britain, the yard, the avoirdupois pound, the troy pound, and the apothecaries pound are identical with the units of the same names used in the United States. The tables of British linear measure, troy mass, and apothecaries mass are the same as the corresponding United States tables, except for the British spelling "drachm" in the table of apothecaries mass. The table of British avoirdupois mass is the same as the United States table up to 1 pound; above that point the table reads:

14 pounds	= 1 stone	
2 stones	= 1 quarter	= 28 pounds
4 quarters	= 1 hundredweight	= 112 pounds
20 hundredweight	= 1 ton	= 2,240 pounds

The present British gallon and bushel—known as the "Imperial gallon" and "Imperial bushel"—are, respectively, about 20 percent and 3 percent larger than the United States gallon and bushel. The Imperial gallon is defined as the volume of 10 avoirdupois pounds of water under specified conditions, and the Imperial bushel is defined as 8 Imperial gallons. Also, the subdivision of the Imperial gallon as presented in the table of British apothecaries fluid measure differs in two important respects from the corresponding United States sub-division, in that the Imperial gallon is divided into 160 fluid ounces (whereas the United States gallon is divided into 128 fluid ounces), and a "fluid scruple" is included. The full table of British measures of capacity (which are used alike for liquid and for dry commodities) is as follows:

4 gills	= 1 pint
2 pints	= 1 quart
4 quarts	= 1 gallon
2 gallons	= 1 peck
8 gallons (4 pecks)	= 1 bushel
8 bushels	= 1 quarter

The full table of British apothecaries measure is as follows:

20 minims	= 1 fluid scruple
3 fluid scruples	= 1 fluid drachm
	= 60 minims
8 fluid drachm	= 1 fluid ounce
20 fluid ounces	= 1 pint
8 pints	= 1 gallon (160 fluid ounces)

NOTES

Figures at the start of each paragraph refer to page numbers.

INTRODUCTION

1–2 The description of East Liverpool and the Ohio Valley is based on personal observation. The East Liverpool museum quotes the town's nineteenth-century claim to have been "the pottery capital of the world." Hutchins's soil description is from his first report to Congress.

2 "For the distance of 46 chains . . .": Thomas Hutchins, "A brief account of the soil and timber in that part of the Western Territory through which and [*sic*] East-West line has been surveyed agreeable to an Ordinance of Congress of the 20th May 1785," *Papers of the Continental Congress* 60: 182–83.

2 Greek historians, among them Proclus Diadochus in the fifth century c.e. (*Commentary on Euclid's Elements*), were the first to trace the origin of surveying back to the Nile's inundations, and to suggest that its practices were brought to Greece by the geometer Thales of Miletus in the sixth century b.c.e., to be broadened into theory by Pythagoras some time later, and eventually refined into the theorems of Euclid in the fourth century b.c.e. Twentieth-century archaeology suggested an independent, possibly earlier origin in the annually flooded valleys of the Tigris-Euphrates in present-day Iraq.

4 The study of ancient measures, or metrology, is a minefield of learning and conjecture that sometimes rivals astrology in its arcane mysteries. (For a classic study, see Algernon E. Berriman, *Historical Metrology* [London: J. M. Dent, 1953].) The conjectural part of metrology depends on an assumption that prehistoric measures, once established, did not vary, an assumption that cannot be sustained once history begins. My own belief is that variability was the norm, and that where prehistoric measures can be shown to be standardized, such as the Babylonian and Egyptian cubit, or simply to have appeared, such as the "megalithic yard" that Alexander Thom detected at Stonehenge in the 1960s, their application was for a specific purpose. The "royal" cubit holds good for measuring the Great Pyramid, but different cubits are used elsewhere; and the megalithic yard works for Stonehenge and other stone circles but does not otherwise reappear with any exactness. It is significant that of the two surviving examples of the oldest weight, the Babylonian mina, one weighs 1.4 pounds and the other approximately 50 percent more.

The sites in central and southeastern Anatolia, modern Turkey, date from between 8500 b.c.e. for Hallan Çemi Tepesi to 4500 b.c.e. for Çatal Hüyük, and have produced a range of increasingly sophisticated artefacts of exchange and social life. Measurement appears even in the earliest of these sites in the form of a series of uniform batons made of soft stone and incised with various numbers of notches. (M. H. Gates, "Archeology in Turkey," *American Journal of Archaeology* 98, no. 2 [1994]: 249–78.)

4 The evidence for a change in landownership occupies much of the next two chapters.

5 Gunter's chain became the standard surveyor's instrument, but it was not the only chain to appear in the seventeenth century. Aaron Rathborne was the first to devise a 22-yard chain, but it lacked the decimal division that gave Gunter's its superiority. Vincent Wing subsequently invented an 80-link chain that he claimed was more suited to 4-based area measurements. (A. W. Richeson, *English Land Measuring to 1880* [Cambridge, Mass.: MIT Press, 1966].) In 1785 William Roy used a 100-foot chain at the start of the Ordnance Survey of Britain, and in Texas the usual chain was based on the *vara*. None of them, however, was used as widely as Gunter's. Long after its replacement in the mid–twentieth century, the thin steel band introduced in its place was still referred to as a chain and divided into 100 links, and today's surveyors still routinely translate the metric measurements from global positioning systems and laser-beam instruments into Gunter's language.

6 Evidence of the remarkable unanimity among so many of the founding fathers about the need for reform of American weights and measures appears in chapters 8–9.

ONE. THE INVENTION OF LANDED PROPERTY

Sources for the background of inflation and agrarian change in Tudor England include G. R. Elton, *England Under the Tudors* (London: Methuen, 1974), and Penry Williams, *Life in Tudor England* (London: Batsford, 1964). From the 1980s, the burgeoning history of cartography has transformed our understanding of what mapmaking implied and has given rise to a series of new writing on the changing nature of landownership in the sixteenth and seventeenth centuries. Among these latter sources are: T. H. Aston and C. H. E. Philpin, eds., *The Brenner Debate: Agrarian Class Structure and Economic Development in Pre-Industrial Europe* (Cambridge: Cambridge University Press, 1987); David Buisseret, ed., *Monarchs, Ministers and Maps: The Emergence of Cartography as a Tool of Government in Early Modern Europe* (Chicago: University of Chicago Press, 1992); and Andrew McRae, *God Speed the Plough: The Representation of Agrarian England, 1500–1660* (Cambridge: Cambridge University Press, 1996).

7 John Fitzherbert, *The Art of Husbandry* (London, 1523), and A. W. Richeson, *English Land Measuring to 1880* (Cambridge, Mass: MIT Press, 1966).

8 Sir Thomas Elyot, *The Boke Named the Governour* (London, 1531).

9–10 In German-speaking Europe, surveying manuals were available by the early seventeenth century, but estate mapping was negligible. (David Buisseret, "Estate Maps in the Old World," in David Buisseret, ed., *Rural Images: Estate Maps in the Old and New Worlds* [Chicago: University of Chicago Press, 1996].)

In Scandinavia, Sweden's first national map appeared in the sixteenth century, but local maps did not appear for another century. (Ibid.)

In France, Marc Bloch observed in 1929 that there seemed to be no seigneurial *plans parcellaires* before 1650, and very few in the century after that. (Cited in ibid.) In Spain, the Casa de Contrataciøn in Seville began to compile a world map in 1508, but there would be no estate maps until the eighteenth century. (Barbara E. Mundy, *The Mapping of New Spain: Indigenous Cartography and the Maps of the Relaciones Geograficas* [Chicago: University of Chicago Press, 1996].)

10 "But in England . . .": Cited in Swen Voekel. "Upon the Juddaine View: State, Civil Society, and Surveillance in Early Modern England." *Early Modern Literary Studies,* Sept. 1998.

10 "a black coffer . . .": Peter Barber, "England II: Monarchs, Ministers, and Maps, 1550–1625," in Buisseret, *Monarchs, Ministers, and Maps;* cited in Voekel, "'Upon the Suddaine View.'"

10 Richard Benese, *This boke sheweth the maner of measurynge of all maner of lande* . . . (London, 1538).

10 For variable units in Domesday Book, see F. W. Maitland, *Domesday Book and Beyond: Three Essays in the Early History of England* (Cambridge, 1897).

12 For John Palmer's story, see Penry Williams, *Life in Tudor England* (London: Batsford, 1964).

13 Robert Recorde's verse: Andrew McRae, *God Speed the Plough: The Representation of Agrarian England, 1500–1660* (Cambridge: Cambridge University Press, 1996).

13–15 The scanty details of Edmund Gunter's life are to be found in Richeson, *English Land Measuring to 1880; Dictionary of National Biography* (Oxford: Oxford University Press, 1949–50); E. G. R. Taylor, *The Mathematical Practitioners of Tudor and Stuart England* (Cambridge: Cambridge University Press, 1954); Charles H. Cotter, "Edmund Gunter (1581–1626)," *Journal of Navigation* 34 (1981); manuscript in possession of Gresham College, City of London: the Oxford gossip and Savile story come from John Aubrey's *Brief Lives* (London: Cresset Press, 1949), pp. 131–32.

16 "He did open men's understandings . . .": John Aubrey, *Brief Lives.*

19 "found the King's Mannors . . .": Daniel W. Hollis, "The Crown Lands and the Financial Dilemma in Stuart England," *Albion* 26, no. 3 (1994): cited in Voekel, "'Upon the Suddaine View.'"

TWO. PRECISE CONFUSION

The major authority for the lore of premetric measures, especially their variability, is Witold Kula (trans. R. Szreter), *Measures and Men* (Princeton, N.J.: Princeton University Press, 1986). The postmetric history is well covered in Ronald Edward Zupko, *Revolution in Measurement: Western European Weights and Measures Since the Age of Science* (Philadelphia: American Philosophical Society, 1990). A more general approach comes in Arthur Klein, *The World of Measurements* (London: Allen and Unwin, 1975).

21–22 Variable measures: Kula, *Measures and Men.*

23 "Four fingers make one palm . . .": Quoted by Leonardo da Vinci in 1490 captioning of his drawing *Proportions of the Human Body After Vitruvius.*

23–24 French weights and measures: Kula, *Measures and Men.*

24 "their exact marking . . .": Josiah Child, *Brief Observations Concerning Trade and Interest of Money* (London, 1688).

25 "called forth by the uncertainty. . . .": Cited in Klein, *The World of Measurements*.

26 "Foure graines . . .": Cited in ibid.

28 "As for the Natiues . . .": Winthrop's comment on Roger Williams's ideas, December 27, 1633. *The journal of John Winthrop, 1630–1649,* edited by Richard S. Dunn, James Savage, and Laetitia Yeandle (Cambridge, Mass. and London: The Belknap Press of Harvard University Press, 1996).

THREE. WHO OWNED AMERICA?

The main sources for American land distribution are D. W. Meinig, *The Shaping of America: A Geographical Perspective on 500 Years of History,* vols. 1–2 (New Haven, Conn.: Yale University Press, 1986, 1993); Thomas Abernethy, *Western Lands and the American Revolution* (Appleton: Century, 1937); Michael P. Conzen, ed., *The Making of the American Landscape*: Unwin Hyman, 1990); Edward T. Price, *Dividing the Land: Early American Beginnings of Our Private Property Mosaic* (Chicago: University of Chicago Press, 1995); and John Stilgoe, *The Common Landscape of America, 1580–1845* (New Haven, Conn.: Yale University Press, 1982).

29 "We stood a while . . .": John Bereton, "Briefe and True Relation of the Discovery of the North Part of Virginia, 1602," cited in H. S. Burrage, ed., *Early English and French Voyages* (New York, 1906).

29 "The mildnesse of the aire . . .": John Smith, "Description of Virginia and Proceedings of the Colonie By Captain John Smith, 1612," in Tyler, Lyon G., ed, *Narratives of Early Virginia 1606–1625.* (New York: Charles Scribner's Sons, 1907), pp. 97–98.

29 "I will end therefore . . .": Andrew White, "A Briefe Relation of the Voyage unto Maryland, 1634," cited in Gerald L. Smith, *God and the Land: Natural Theology and Natural History* (http://smith2.sewanee.edu/gsmith/Texts/Ecology/GodAndTheLand.html).

31–32 "The Reeds which grew . . .": William Byrd II, *The History of the Dividing Line* (Richmond, Va., 1866).

32 "Extraordinary fatigue . . ."; "the rains, the hot weather and the insects . . .": Both cited in Louise Pettus, *Surveying the Border Between the Carolinas* (http://www.surveyhistory.org/south_carolina_early_survey_history.htm).

33 Mason-Dixon line: *Encyclopaedia Britannica;* Silvio Bedini, *Thinkers and Tinkers.* (New York: Scribner's, 1975), p. 139. Thomas Pynchon, *Mason and Dixon* (New York: Holt, 1997).

34 "A certain quantity of land . . ."; "so that the neighbor . . .": William Penn, *Concessions to the Province of Pennsylvania,* (1681).

34 "by lines running East . . .": *Fundamental Constitutions of Carolina* (1669).

34 Georgia's and Savannah's squares: John Reps, *The Making of Urban America: A History of City Planning in the United States* (Princeton, N.J.: Princeton University Press, 1965).

35 "And so [I] assigned to every family . . .": William Bradford, *History of Plymouth Plantation* (c. 1650).

35 "How many men have since coveted . . .": Cited in John Stilgoe, *The Common Language of America 1580–1845* (New Haven, Conn.: Yale University Press, 1982).

36 "In Virginia . . .": Cited in "Virginia Timeline," Library of Congress http://memory.loc.gov/ammem/mtjhtml/mtjvatm4.html.

36–37 "it was against the laws of God . . .": Cited in D. W. Meinig, *The Shaping of America: A Geographical Perspective on 500 Years of History,* vol. 2 (New Haven, Conn.: Yale University Press, 1993).

38 "[Descending in the dark from a mountain], we fell into a place . . .": Thomas Lewis, *The Fairfax Line: Thomas Lewis's Journal of 1746* (New Market, Va., 1925).

38 "full of rocks and cavities . . .": ibid.

39 "A doubloon is my constant gain . . .": John C. Fitzpatrick, ed., *Writings of George Washington from the Original Manuscript Sources, 1745–1799* (Washington, D.C.: Government Printing Office, 1931–44).

39 The breakdown of the Carolina survey: Louise Pettus, *Surveying the Border Between the Carolinas.*

40 Virginia "headrights" measurement and "metes and bounds": Abernethy, *Western Lands;* Conzen, *The Making of the American Landscape.*

40–41 New England land distribution: Stilgoe, *The Common Landscape of America; Dividing the Land.*

41 "Northward lys the lott . . .": From the *New England Historical and Genealogical Register* 12, no. 4 (1859).

41–44 Comparison of Spanish, French, and British American land distribution: Meinig, *The Shaping of America.*

45 "The greatest Estates . . .": Cited in Charles Royster, *The Fabulous History of the Dismal Swamp Company: A Story of George Washington's Times* (New York: Knopf, 2000).

45 "Their vallies are . . .": Cited in Smith, *God and the Land.*

46 Hutchins's map: Frederick C. Hicks, ed., *Thomas Hutchins: A Topographical Description of Va, Pa, Md + No Carolina* (London, 1778; reprint, Cleveland: Burrows Bros., 1904).

47 "The country might invite a prince from his palace . . .": Richard Henderson's advertisement. Cited in the frontispiece of *The Conquest of the Old Southwest. The romantic story of the early pioneers into Virginia, the Carolinas, Tennessee, and Kentucky, 1740–1790* by Archibald Henderson (New York: Century, 1920).

FOUR. LIFE, LIBERTY, OR WHAT?

The activities of American land companies are covered primarily in Thomas Abernethy, *Western Lands and the American Revolution* (Boston: Appleton Century, 1937); Shaw Livermore, *Early American Land Companies and Their Influence, a Corporate Development* (Commonwealth Fund, 1939); A. M. Sakolski, *The Great American Land Bubble: The Amazing Story of Land-Grabbing, Speculations, and Booms from Colonial Days to the Present Time* (New York and London: Harper and Brothers, 1932).

The prime source for Thomas Jefferson is *The Papers of Thomas Jefferson,* edited by Julian Boyd and published by the Princeton University Press. Of these papers, I have relied chiefly on his letters, autobiography, and *Notes on the State of Virginia,* usually supplemented by online texts supplied by the Avalon Project. The biography of Jefferson I have relied on most heavily is Joseph Ellis's *American Sphinx* (New York: Knopf, 1997), supported by William H. Adams, *The Paris Years of Thomas Jefferson* (New Haven, Conn.: Yale University Press, 1997); and Anthony F. C. Wallace, *Jefferson and the Indians: The Tragic Fate of the First Americans* (Cambridge, Mass. and London: The Belknap Press of Harvard University Press, 1999).

49 "I can never look upon . . .": Cited in Edward Redmond, "George Washington: Surveyor and Mapmaker," Library of Congress, no date.

49–50 Crawford's killing: *Virginia Gazette,* August 3, 1782; cited in Hazel Dicken-Garcia, *To Western Woods* (Rutherford, N.J.: Associated University Press, 1991).

50 "who maintained that to kill an Indian . . .": Cited in Anthony Wallace, *Jefferson and the Indians: the tragic fate of the first Americans.* 1999.

51 "One half of England . . .": George Croghan writing to Sir William Johnson 1766; cited in A.M. Sakolski, *The Great American Land Bubble* (New York and London: Harper and Brothers, 1932).

51–53 Details of Rufus Putnam's life come from *Memoirs of Rufus Putnam,* ed. Rowena Buell (Boston: Houghton Mifflin, 1904).

53–60 Jefferson's character and actions are derived from the authorities cited in the second paragraph of the notes to this chapter.

54–56 William Small's gift for friendship can be gauged from the grief-stricken letter that the industrialist Matthew Boulton sent to his business partner, James Watt, when Small died in 1775: "Dear Watt, You have lost a friend so have I. Take him all in all we ne'er shall see his like again. My loss is as inexpressible as it is irreparable. I am ready to burst. I can't write more but remain Your inconsolable and affectionate Friend, Matthew Boulton."

FIVE. SIMPLE ARITHMETIC

The sources for land company activities and Jefferson's life are those cited in the first two paragraphs of the notes to chapter 4. The politics of the Continental Congress and early U.S. Congress are covered in Stanley Elkins and Eric MacKitrick, *The Age of Federalism* (Oxford: Oxford University Press, 1995).

63–64 The origins of the Ohio Company of Associates: Thomas H. Smith, *The Mapping of Ohio* (Kent, Ohio: Kent State University Press, 1977); Jeffrey Pasley, "Private Access and Public Power," in *The House and the Senate in the 1790s: Petitioning, Lobbying, and Institutional Development* (Kent, Ohio: Ohio University Press, 2002).

64 "His whole figure . . .": Kenneth R. Bowling and Helen E. Veit, eds., *The Diary of William Maclay and Other Notes on Senate Debates, Documentary History of the First Federal Congress of the United States of America,* vol. 9 (Baltimore, 1988), p. 275.

64–67 Robert Morris: As yet, no biography has been written of this seminal figure, but one must come soon. Financial histories refer to his role as superintendent of finance, legal authorities cite his North American Land Company (founded in 1765 and still in existence) as the first "pure trust" to be established in North America, articles on the tobacco trade allude to his French monopoly, specialists on the Far East mention his trading connections with India and China, but none seems able to explain exactly what he did in any of these spheres. Through the business firm of Willing and Morris, he was linked to banks like Barings in London, to financiers like John Swanwick, and to speculators like William Bingham, and his Bank of North America, founded in 1781, was both the first financial institution chartered by the United States and the center of New York's financial expertise. He was the ghost in America's economic machine.

66 "Every one remembers . . .": Jefferson's *Notes on a Coinage*, Mar/May 1784.

67 "it is happy for us . . .": "Robert Morris to the President of Congress" (1782), in *The Papers of Thomas Jefferson.*

67–68 Evidence for the disorder in American weights and measures occurs in pamphlets such as John Bordley Beale's *On monies, coins, weights, and measures, proposed for the United States of America* (Philadelphia, 1789); proof was provided by research instituted by John Quincy Adams for *A Report upon Weights and Measures* (Washington, D.C., 1821), coupled with Jefferson's references ("Report on Weights and Measures" [1790], in *The Papers of Thomas Jefferson*) to the chaos in British legislation and standards on which American measures depended.

68 The date of "Some Thoughts on a Coinage" (*The Papers of Thomas Jefferson*), once thought to belong to Jefferson's work on weights and measures in 1790, was established by the *Papers'* editor, Julian Boyd. The chronology of political events in 1784 and membership of committees are taken from Boyd's editorial comments.

71 "When I consider . . .": John Adams writing to his wife, Abigail, May 1776, *Familiar Letters of John Adams and his wife Abigail Adams, during the Revolution,* with a memoir of Mrs. Adams, by Charles Francis Adams. (New York: Hurd & Houghton, 1876), pp. xxxii, 424. *The book of Abigail and John: selected letters of the Adams family, 1762–1784,* edited and with an introduction by L. H. Butterfield, Marc Friedlaender, and Mary-Jo Kline. (Cambridge, Mass.: Harvard University Press, 1975), pp. ix, 411, p. [1] leaf of plates.

72 "I am much opposed . . .": Rufus Putnam, *Memoirs of Rufus Putnam,* ed. Rowena Buell (Boston: Houghton Mifflin, 1904).

73 "It deviates I believe. . . .": Cited in W. D. Pattison, *1957—The Beginning of the American Rectangular Land Survey System* (Ohio Historical Society, 1970).

SIX. A LINE DRAWN IN THE WILDERNESS

The early surveying and settlement of Ohio is well covered in Thomas H. Smith's *The Mapping of Ohio* (Kent, Ohio: Kent State University Press, 1977). Thomas Abernethy, *Western Lands and the American Revolution,* and A. M. Sakolski's entertaining *The Great American Land Bubble: The Amazing Story of Land-Grabbing, Speculations, and Booms from Colonial Days to the Present Time* (New York and London: Harper and Brothers, 1932) are useful for the activities of the Ohio Company, Scioto Company, and of John Cleves Symmes.

74–75 "This morning continued the Vista . . .": Andrew Porter's "Journal" is cited in *Pennsylvania Magazine of History and Biography,* vol. 4 (1880).

75–77 Hutchins's life: Thomas Hutchins. Smith, *The Mapping of the Ohio;* Silvio Bedini,

"William Gerard De Brahm, Geographer & Surveyor," *Professional Surveyor,* October 1996, vol. 16, no. 7.

81 Cutler's work as a pioneer lobbyist: Jeffrey Pasley, "Private Access and Public Power," in *The House and the Senate in the 1790s: Petitioning, Lobbying, and Institutional Development* (Ohio University Press, 2002).

82 "There will be one advantage . . .": Cited in Smith, *The Mapping of Ohio.*

83 "to dabble in federal filth": Letter to Madison, July 1791, in *The Papers of Thomas Jefferson.*

83 Sale of Seven Ranges: Michael Flynn, "The Origin and Development of the Rectangular Survey System" (www.landman.org/landman5/flynn.html).

83 "The Land is too rich . . .": Cited in Smith, *The Mapping of Ohio.*

83 "the lands are of so versatile a nature . . .": Cited in ibid.

84 Hutchins's obituaries: Silvio Bedini, *Professional Surveyor.*

85–86 Early plat of Marietta: John Reps, *The Making of Urban America: A History of City Planning in the United States* (Princeton, N.J.: Princeton University Press, 1965).

86 "In a covered Country . . .": Rufus Putnam, *Memoirs of Rufus Putnam,* ed. Rowena Buell (Boston: Houghton Mifflin, 1904).

86–87 Ludlow's wayward surveys: White, *A History of the Rectangular Survey System*; Sakolski, *The Great American Land Bubble.*

86 "scarcely two sections . . .": Albert C. White, *A History of the Rectangular Survey System* (Washington, D.C.: Government Printing Office, 1982).

86 "Before going a mile . . .": C. H. van Orden, cited in W. D. Pattison, 1957, *The Beginning of the American Land Survey System* (Ohio Historical Society, 1970).

87 Symmes and the Miami Purchase: Sakolski, *The Great American Land Bubble;* Smith, *The Mapping of Ohio.*

88 "Your presence here . . .": Cited in Jeffrey Pasley, "Private Access and Public Power." *The House and the Senate in the 1790s: Petitioning, Lobbying, and Institutional Development* (Ohio University Press, 2002).

88 "No colony in America was ever settled . . .": Cited in Henry Howe, *Historical Collections of Ohio* (Norwalk, Ohio: Lanning Printing Co., 1898).

SEVEN. THE FRENCH DIMENSION

For Jefferson's period in France, William H. Adams, *The Paris Years of Thomas Jefferson* (New Haven, Conn.: Yale University Press, 1997), is an excellent source. The scientific background is covered in Tore Frängsmyr, J. L. Heilbron, Robin Rider, et al., *The Quantifying Spirit of the Eighteenth Century* (University of California Press, 1990), and M. Norton Wise, ed., *The Values of Precision* (Princeton, N.J.: Princeton University Press, 1995); Witold Kula's invaluable *Measures and Men* (Princeton, N.J.: Princeton University Press, 1986) provides a European perspective on measures; a specifically French view comes from L. Marquet, A. Le Bouch, and Y. Roussel, *Le Système métrique hier et aujourd'hui* (Amiens: ADCS, 1997), and the catalog to the bicentennial exhibition on the meter at the Centre Nationale des Arts et Métre (Paris: CNAM, 1989). For biographical details of individual scientists, I have used the incomparable Web site of the School of Mathematics and Statistics at the University of St. Andrews in Scotland (www-history.mcs.st-andrews.ac.uk/history).

89–90 The development of Paris: Adams, *The Paris Years of Thomas Jefferson.*

90 Activities of *géomètres:* Ken Alder, "A Revolution to Measure," in *The Values of Precision* (Princeton, N.J.: Princeton University Press, 1995).

90 "The *arpent* is not divisible . . .": Cited in Kula, *Measures and Men.*

91 "The unending proliferation . . .": Cited in ibid.

92 Jefferson's widow: Letter to Madison, 1785, in *The Papers of Thomas Jefferson.*

93 *Cahiers de doléances:* Cited in Kula, *Measures and Men.*

93–96 Condorcet: Adams, *The Paris Years of Thomas Jefferson;* "Condorcet" (www-history.mcs.st-andrews.ac.uk/history).

95–96 "their eminence in science . . .": Thomas Jefferson, Autobiography in *The Papers of Thomas Jefferson.*

96 "That nation hates us . . .": Letter to John Page, March 1786, in *The Papers of Thomas Jefferson.*

96 Jefferson's "Universal Equatorial Instrument": Silvio Bedini, *Professional Surveyor.*

97 Carysfort committee: Arthur Klein, *The World of Measurements* (London: Allen and Unwin, 1975); "Report on Weights and Measures" (1790), in *The Papers of Thomas Jefferson.*

98–100 Origins of British Ordnance Survey: Charles Close, *The Early Years of the Ordnance Survey* (Whitstable, 1969).

100–101 Talleyrand's letter and Riggs Miller's "the poor thresh out corn . . .": Hansard, 1790.

EIGHT. DEMOCRATIC DECIMALS

The dominant source of information on Jefferson's proposals for reform of the United States's weights and measures is *The Papers of Thomas Jefferson* and the authoritative comments of its editor, Julian Boyd. They are supplemented by C. D. Hellman's article "Jefferson's Efforts Towards the Decimalization of United States Weights and Measures," *Isis* 16 (1931).

104 A gallimaufry of measures; *Ruthe* = rod or 16.5 Rhineland feet; Scots mile = 1,984 yards; Irish acre = 7,840 square yards; 32 Scots mutchkins = 16 choppins = 1 Scots gallon of beer (about three times the size of the English gallon); flitch = side of bacon; fardel = large bale of cloth; hattock = 10 sheaves; *Globen* = 10 ends of flax; *Quentchen* = about 4 grains of gold; *Zehnling* = 10 skins. Ronald Edward Zupko, *Revolution in Measurement: Western European Weights and Measures Since the Age of Science* (Philadelphia: American Philosophical Society, 1990).

104 "frauds and deceits": John Quincy Adams, *A Report upon Weights and Measures* (Washington, D.C., 1821).

105 Philadelphia and Maryland measures: John Bordley Beale, *On Monies, Coins, Weights, and Measures, Proposed for the United States of America* (Philadelphia, 1789).

106 "the whole mass of the people . . .": Jefferson's preamble to his "Report on Weights and Measures" (1790), in *The Papers of Thomas Jefferson.*

106–113 Jefferson's attempt to establish a new system of weights and measures: ibid.

113–114 Settlers in Spanish and French America: D. W. Meinig, *The Shaping of America: A Geographical Perspective on 500 Years of History,* (New Haven, Conn.: Yale University

Press); Thomas Abernethy, *Western Lands and the American Revolution* (Appleton Century, 1937).

113–115 Hamilton and Jefferson: Stanley Elkins and Eric MacKitrick, *The Age of Federalism* (Oxford: Oxford University Press, 1995).

116 Public opinion on weights and measures: Julian Boyd, in *The Papers of Thomas Jefferson.*

NINE. THE BIRTH OF THE METRIC SYSTEM

The sources for this chapter are largely those for chapter 7.

117–118 Putnam's attitude: Rufus Putnam, *Memoirs of Rufus Putnam*, ed. Rowena Buell (Boston: Houghton Mifflin, 1904).

118 Senate proceedings on weights and measures: December 28, 1790, first Izard committee appointed to consider Jefferson's report; March 1, 1791, first Izard committee report delivered; November 1, 1791, second Izard committee appointed; April 5, 1792, second Izard committee report delivered; report considered December 3 and 17, 1792; debated December 18, 1792; revised January 29, 1793—see *Gazette of the United States,* March 13, 1793, considered February 8, 1793; report referred to Harrison committee December 24, 1795, *Annals of Congress.*

121 John Swanwick's position: *Annals of Congress*, 4th Congress, Session 1.

121–23 French commission's report: Borda, Lagrange, LaPlace, Monge, and Condorcet, *Rapport fait à L'Académie des Sciences sur le choix d'une unité de mesures* (Paris, 1791).

123–24 Louis XVI and Borda: Perre Aubert, "Borda et le système métrique," *Bullétin de Société Borda* 456 (2000).

127 Senate proceedings: See references for p. 118.

128–30 Evolution of the meter: Ken Alder, "A Revolution to Measure," in *The Values of Precision* (Princeton, N.J.: Princeton University Press, 1995); Witold Kula, *Measures and Men* (Princeton, N.J.: Princeton University Press, 1986); and L. Marquet, A. Le Bouch, and Y. Roussel, *Le Système Métrique hier et aujourd'hui* (Amiens: ADCS, 1997).

130 "Surely it demonstrates . . .": Committee of Public Instruction. Cited in Witold Kula, *Measures and Men.*

130–32 Joseph Dombey: E. Hamy, *Joseph Dombey*; E. Dubois, *Le Naturaliste Joseph Dombey* (Bourg-en-Bresse, 1934); manuscript in possession of Museum Nationale d'Histoire Naturelle, Paris).

TEN. DOMBEY'S LUCK

The tragic quality of Dombey's life struck both of his biographers, E. Hamy and E. Dubois (see references for pp. 153–55), but each was concerned with his career as a botanist rather than as the bearer of the metric system.

133–38 Dombey's fate: See references for pp. 153–55.

138–40 House of Representatives debates on weights and measures: *Annals of Congress,* 4th Congress, cols. 1376–83.

140–41 Wayne and the Battle of Fallen Timbers: Anthony F. C. Wallace, *Jefferson and the*

Indians: The Tragic Fate of the First Americans (London: Belknap, 1999); John Selby, *The Conquest of the American West* (London: Allen and Unwin, 1975).

142 "This specifically calls for ...": Albert White, *A History of the Rectangular Survey,* p. 29.

ELEVEN. THE END OF PUTNAM

The authority on the origins of the public lands survey is W. D. Pattison, *The Beginning of the American Rectangular Land Survey System* (Ohio Historical Society, 1970). C. Albert White's *A History of the Rectangular Survey System* (Washington, D.C.: Government Printing Office, 1982) is the essential guide to the public lands survey in the nineteenth century.

143 "the emigration to this country ...": Cited in A. M. Sakolski, *The Great American Land Bubble: The Amazing Story of Land-Grabbing, Speculations, and Booms from Colonial Days to the Present Time* (New York: Harper and Brothers, 1932).

143–44 Putnam's house: Personal observation by the author.

144 Ludlow's surveys: White, *A History of the Rectangular Survey System;* Sakolski, *The Great American Land Bubble.*

145 Virginia Military Reserve: White, *A History of the Rectangular Survey System.*

145–47 Cleveland's streets: Thomas H. Smith, *The Mapping of Ohio* (Kent, Ohio: Kent State University Press, 1977).

147–48 Putnam's dismissal: Rufus Putnam, *Memoirs of Rufus Putnam,* ed. Rowena Buell (Boston: Houghton Mifflin, 1904).

148–150 Land speculation: Thomas Abernethy, *Western Lands and the American Revolution* (Boston: Appleton Century, 1937); Sakolski, *The Great American Land Bubble.*

152–53 This passage on southern Appalachian land speculation leans heavily on Wilma A. Dunaway, "Speculators and Settler Capitalists," in *Appalachia in the Making: The Mountain South in the Nineteenth Century* (Chapel Hill, N.C.: University of North Carolina Press, 1995).

154–55 Yazoo land fraud: Abernethy, *Western Lands and the American Revolution;* Sakolski, *The Great American Land Bubble.*

154 Georgia compared to Russia: Christian Lucky, "Public Theft in Early America and Contemporary Russia," *East European Constitutional Review,* fall 1997.

155–56 Briggs' problems: White, *A History of the Rectangular Survey System.*

156–57 Talleyrand and Morris: William H. Adams, *The Paris Years of Thomas Jefferson* (New Haven, Conn.: Yale University Press).

157 Talleyrand in Maine: Simon Schama, *Citizens: A Chronicle of the French Revolution* (New York: Vintage Books, 1990).

157–59 Louisiana Purchase: Abernethy, *Western Lands and the American Revolution.*

158–59 Silas Bent: White, *A History of the Rectangular Survey System.*

TWELVE. THE IMMACULATE GRID

160 Jared Mansfield: Silvio Bedini, *Professional Surveyor.*

161 Route 277: Personal observation by the author.

160–70 Survey: C. Albert White, *A History of the Rectangular Survey System* (Washington, D.C.: Government Printing Office, 1982); W. D. Pattison, *The Beginning of the American Rectangular Land Survey System* (Ohio Historical Society, 1970).

163 Land prices: A. M. Sakolski, *The Great American Land Bubble: The Amazing Story of Land-Grabbing, Speculation, and Booms from Colonial Days to the Present Time.* (New York and London: Harper and Brothers, 1932).

163–64 John Pulliam: John Mack Faragher, *Sugar Creek: Life on the Illinois Prairie,* (New Haven: Conn.: Yale University Press, 1986).

166–68 Surveyor's notes: White, *A History of the Rectangular Survey System.*

168–69 Iowa squatter: Roscoe L. Lokken, *Iowa Public Land Disposal* (Iowa City: State Historical Society of Iowa, 1942).

170 Alabama land office: Wilma A. Dunaway, "Speculators and Settler Capitalists," in *Appalachia in the Making: The Mountain South in the Nineteenth Century* (Chapel Hill, N.C.: University of North Carolina Press, 1995).

170 "For actual property . . .": Cited in Joseph Ellis, *American Sphinx* (New York: Knopf, 1997).

171 "The possession of land . . .": Harriet Martineau, *Society in America* (London, 1837).

171 "Every industrious citizen . . ."; "The land being purely his own . . .": John Melish, *Travels in the United States 1806, 1807, and 1809* (Philadelphia, 1812).

172 "I *own* here a far better estate . . .": Morris Birkbeck, *Letters from Illinois,* 1817.

172 "As soon as the government . . .": Frederick Marryat, *A Diary in America,* ed. Sydney Jackman (New York: Knopf, 1962).

173 "They pay neither taxes nor tythes . . ."; "Any man's son . . .": Frances Trollope, *Domestic Manners of the Americans* (1832; London: Imprint Society, 1969).

174 "The frontier promoted . . .": Frederick J. Turner, *The Frontier in American History* (New York: Dover Publications, 1996).

THIRTEEN. THE SHAPE OF CITIES

The major source of information for this chapter is John Reps's magical *The Making of Urban America: A History of City Planning in the United States* (Princeton, N.J.: Princeton University Press, 1965).

176–79 The expansion of Manhattan: Charles Lockwood, *Manhattan Moves Uptown: An Illustrated History* (Boston: Houghton Mifflin, 1976).

177 "Whether we should confine ourselves . . ."; "In considering . . .": Reps. *The Making of Urban America.*

179 "See what I intend . . .": Lockwood, *Manhattan Moves Uptown.*

180 "needs a space of ground . . .": Frederick Law Olmsted, cited in Reps, *The Making of Urban America.*

180 New York City tenements: Jacob A. Riis, *How the Other Half Lives: Studies Among the Tenements of New York* (New York: Charles Scribner's Sons, 1890).

180 "The fact that it was . . .": Reps, *The Making of Urban America.*

181 "The streets were crowded . . .": Harriet Martineau, *Society in America* (London, 1837).

181–82 "We send by this mail . . .": Cited in Reps, *The Making of Urban America.*

184 "An office boy can figure . . .": Lewis Mumford, *The Culture of Cities* (San Diego, Calif.: Harcourt, 1970).

184–85 Railroad development: Reps, *The Making of Urban America.*

FOURTEEN. HASSLER'S PASSION

The principal source of information for Ferdinand Hassler's life is Florian Cajori's *The Chequered Career of Ferdinand Rudolph Hassler* (Boston: Christopher Publishing House, 1929). Hassler's own writings on both the coastal survey and weights and measures are gathered in his *Principal Documents Relating to the Survey of the Coast* (New York, 1834).

188–191 Hassler's career: Cajori, *The Chequered Career.*

191 London instrument makers: Julian Holland, "Pioneer of Precision: Captain Henry Kater," Macleay Museum, 1997. URL-http://www.usyd.edu.au/su/macleay/kater1.html

192–94 Details on Hassler come from Cajori, *The Chequered Career,* and Hassler's *Principal Documents.*

195 "I find every one comprehends . . .": Autobiography, in *The Papers of Thomas Jefferson.*

195–97 Adams's report: John Quincy Adams, *A Report upon Weights and Measures* (Washington, D.C., 1821).

197 "The ancient inhabitants . . .": Bouchon's testimony cited in John Quincy Adams's *A report upon Weights and Measures.*

198–205 Hassler's career: Cajori, *The Chequered Career.*

198 "as an organ of communication . . .": Ric Burns and James Sanders, *New York: An Illustrated History* (New York: Knopf, 1999).

202 "depends on the sharpness of the knife edges . . .": F. R. Hassler, *Comparison of Weights and Measures of Length and Capacity* (Washington, D.C., 1832).

202 "It is believed, however . . .": Cited in "Weights and Measures History" by Arizona Weights and Measures Department. URL http://www.weights.az.gov/MoreAboutUs/History/HistoryT.htm

202 "under the immediate . . .": Report of treasury secretary in documents relating to the construction of uniform standards of weights and measures for the United States from 1832 to 1835.

203–205 American Customary System: Arthur Klein, *The World of Measurements* (London: Allen and Unwin, 1975); R. W. Smith, *The Federal Basis for Weights and Measures. A Historical Review of Federal Legislative Effort, Statutes, and Administrative Action in the Field of Weights and Measures in the United States* (Washington, D.C., National Bureau of Standards, 1958).

205 Hassler's death and stories about him: Cajori, *The Chequered Career.*

FIFTEEN. THE DISPOSSESSED

Details of William Burt's life and quotations are from John Burt's biography *They Left Their Mark: William Austin Burt and His Sons, Surveyors of the Public Domain* (Landmark, 1987).

206–208 Burt in London: Ibid.

209 "Republican Empire of North America": William Gilpin, *The Central Gold Region, the Grain, Pastoral, and Gold Regions of North America. With Some Views of Its Physical Geography, Etc.* (Philadelphia, 1860).

209 Potter's land: Cited in Lane J. Bouman, "The Location and Survey of Oregon Donation Land Claims," URL http://www.plso.org/readingroom/OregonDLC-Bouman.htm

212 "No post set nor bearing taken . . .": Manuscript held in the state archives at the Michigan Historical Center, Lansing.

212 "Dear Companion,": Burt, *They Left Their Mark.*

212 Michigan scenery: Personal observation by the author.

213 "There are but two means . . .": Cited in Anthony F. C. Wallace, *Jefferson and the Indians: The Tragic Fate of the first Americans* (Cambridge, Mass. and London: The Belknap Press of the Harvard University Press, 1999).

213–16 Indian treaties: Charles Kappler, ed., *Indian Affairs. Laws and Treaties* (Washington, D.C., 1904).

215 "They say that when . . .": Cited in Wallace, *Jefferson and the Indians.*

216 The different versions of Seattle's speech are quoted in David M. Buerge's Internet essay "Chief Seattle and Chief Joseph: from Indians to Icon," URL http://content. lib.washington.edu/aipnw/buerge2/buerge2.html.

217 "I love that land . . .": Cited in James Hunter, *A Dance Called America* (Mainstream, 1994).

217 "I am a Kickapoo": Cited in Wallace, *Jefferson and the Indians.*

218 "The race was not over . . .": Cited in Vernon L. Parrington, "The American Scene," in *Main Currents in American Thought* (New York: Harcourt Brace, 1927).

SIXTEEN. THE LIMIT OF ENCLOSURE

219 "There is no long measure . . .": Josiah A. Gregg, *Commerce of the Prairies* (New York: 1844–45).

219–20 Stephen Austin: Edward T. Price, *Dividing the Land: Early American Beginnings of Our Private Property Mosaic* (Chicago: University of Chicago Press, 1995).

221 "Certain title to the land . . .": Cited in Albert C. White, *A History of the Rectangular Survey System* (Washington, D.C.: Government Printing Office, 1982).

221 "From the top of this mountain . . ."; "These valleys and the ravines . . .": Manuscript held by the archives of the Bureau of Land Management, Sacramento, California.

222 "There are many settlers . . .": Manuscript held in the state archives of the Golden State Museum, Sacramento.

222–23 Los Angeles geography: Personal observation by the author.

223 View from aircraft: Ibid.

225 Côteau des Prairies: Ibid.

226 "I purchased in 1845 property . . .": Cited in A. M. Sakolski, *The Great American Land Bubble: The Amazing Story of Land-Grabbing, Speculation, and Booms from Colonial Days to the Present Time* (New York: Harper and Brothers, 1932).

227 Pulliam and Stoute: John Mack Faragher, *Sugar Creek: Life on the Illinois Prairie* (New Haven, Conn.: Yale University Press, 1986).

227 Railroad dream: Jonathan Raban, *Bad Land: An American Romance* (Picador, 1996).

227–28 Railroad advertisements: Posters in Michigan Historical Center, Lansing; North Dakota Institute for Regional Studies, Fargo; South Dakota Cultural Heritage Center, Pierre; etc.

229–30 Grace Fairchild: Walker D. Wyman, *Frontier Woman: The Life of a Woman Homesteader on the Dakota Frontier* (River Falls, Wis.: University of Wisconsin-River Falls Press, 1972).

233 Private property: Harvey M. Jacobs, ed., *Who Owns America? Social Conflict over Property Rights* (Madison: University of Wisconsin Press, 1998).

233 "There is as yet . . .": Aldo Leopold, *A Sand County Almanac* (Oxford University Press, 1987).

234 "The magnitude of the greatest land-measurement project . . .": Hildegard Binder Johnson, *Order upon the Land: The US Rectangular Survey and the Upper Mississippi Country* (London: Oxford University Press, 1976).

235 "up to and including . . .": 1890 report cited in Frederick J. Turner, *The Frontier in American History,* (New York; London: Dover, 1996).

SEVENTEEN. FOUR AGAINST TEN

236–38 For the growth in precision, see M. Norton Wise, ed., *The Values of Precision* (Princeton, N.J.: Princeton University Press, 1995).

237–38 "Although the electrical . . .": Cited in ibid.

238–42 Metric history: Witold Kula (trans. R. Szreter), *Measures and Men* (Princeton, N.J.: Princeton University Press, 1986).

242–43 "When you can measure . . .": Lord Kelvin, "Electrical Units of Measurement," lecture delivered to the Institution of Civil Engineers, 1833.

242–47 Development of the *Système Internationale: The World of Measurements* (London: Allen and Unwin, 1975); Arthur Klein, Web site of the International Bureau of Weights & Measure (http://www.bipm.fr).

EIGHTEEN. METRIC TRIUMPHANT

In addition to government accounts, such as National Institute of Standards and Technology, *Toward a Metric America* (Washington, D.C.: Department of Commerce, 1971), and the British Metrication Board's *Final Report* (London, 1980), which have their own agenda, pro- and anti-metric Web sites carry intriguing, though not necessarily reliable, information; see, for example, http://www.pueblo.gsa.gov/cic—text/misc/usmetric/metric.htm (pro-) and http://www.tysknews.com/Depts/Metrication/metrication.htm (anti-).

248–49 Spread of the metric system: Witold Kula (trans. R. Szreter), *Measures and Men* (Princeton, N.J.: Princeton University Press, 1986), and the Web site http//www.bipm.fr.

252 U.S. metric development: NIST, *Toward a Metric America,* Metric Program, NIST

Web site http://ts.nist.gov/ts/htdocs/200/202/mp0_home.htm.

250–51 U.K. metric development: Metrication Board, *Final Report.* Web sites http://www.footrule.org/ and http://www.bwmaOnline.com.

253 Traditional units in Europe: www.bwmaOnline.com.

254 Canadian metric development: www.ukmetrication.com/canada.htm.

254 "the produce clerks speak . . .": *Washington Post,* May 22, 2000.

254–56 metric confusion: www.bwmaOnline.com.

256–57 Mars Climate Orbiter. *New York Times,* September 23, 24, 30, October 1, November 10, 1999; MCO official Web site, http://www.mars.jpl.nasa/gov/msp98/orbiter/.

258 "The American style . . .": Edward Tenner, quoted in "Waits and Measures," *Mother Jones,* January–February 1999.

EPILOGUE: THE WITNESS TREE

259–60 Lance Bishop: Interview by the author, October 2000.

260 "the measurements of every plot . . .": Response by the Franklin Institute to the United States becoming a signatory to the Treaty of the Meter in 1875; cited in *The Bulletin of the American Iron and Steel Association,* c. 1893.

260 "Nothing can be more perfect . . .": Coleman Sellers, former weigh-master at the Globe Rolling Mills in Cincinnati, 1875; cited in *The Bulletin of the American Iron and Steel Association,* c. 1893.

261–62 Ed Patton: Interview by the author, October 2000.

263 "The Gift Outright": Robert Frost, *A Witness Tree* (New York: Henry Holt, 1942).

BIBLIOGRAPHY

Abernethy, Thomas. *Western Lands and the American Revolution.* Boston: Appleton Century, 1937.

Adams, John Quincy. *A Report upon Weights and Measures.* Washington, D.C., 1821.

Atran, Scott. *The Cognitive Foundations of Natural History.* Cambridge University Press, 1990.

Aubrey, John. *Brief Lives.* London: Cresset Press, 1949.

Beale, John Bordley. *On monies, coins, weights, and measures, proposed for the United States of America.* Philadelphia, 1789.

Bedini, Silvio. "Captain Thomas Hutchins," Parts I and II, *Professional Surveyor,* July/August 1998, vol. 18 no. 5; September 1998, vol. 18, no. 6.

Benese, Richard. *This boke sheweth the maner of measurynge of all maner of lande as woodlande as of lande in the felde and comptynge the true nombre of acres of the same & Newlye invented and complyed.* Southwarke: James Nicolson, 1538.

Bureau of Land Management. *Manual of Instructions for the Survey of Public Land.* Washington, D.C., 1947.

Cajori, Florian. *The Chequered Career of Ferdinand Rudolph Hassler, First Superintendent of the United States Coast Survey. A Chapter in the History of Science in America.* Boston: Christopher Publishing House, 1929.

Cocker, Edward. *Cocker's Decimal Arithmetic.* 37th ed. London, 1720.

Conzen, Michael P., ed. *The Making of the American Landscape.* London: Unwin Hyman, 1990.

Digges, Leonard. *A Booke named Tectonicon briefely shewing the exact measuring, and speedie reckoning of all maner of Land, Squares, Timber, Stone, Steeples, Pillers, Globes, etc.* London, 1556.

Dubois, E. *Le Naturaliste Joseph Dombey.* Bourg-en-Bresse, 1934.

Dunaway, Wilma A. "Speculators and Settler Capitalists." In *Appalachia in the Making: The Mountain South in the Nineteenth Century.* Chapel Hill, N.C.: University of North Carolina Press, 1995.

Durham, Philip, and Jones, Everett, ed. *The Frontier in American Literature.* New York: Odyssey Press, 1969.

Elkins, Stanley, and MacKitrick, Eric. *The Age of Federalism.* Oxford: Oxford University Press, 1995.

Elton, G. R. *England Under the Tudors.* London: Methuen, 1974.

Ferguson, Kitty. *Measuring the Universe.* New York: Walker, 1999.

Fitzherbert, John. *The Art of Husbandry.* London, 1523.

Frängsmyr, Tore; Heilbron, J. L., Rider, Robin, et al. *The Quantifying Spirit of the Eighteenth Century.* Berkeley: University of California Press, 1990.

Gibson, Robert. *A Treatise of Practical Surveying: With Alterations and Amendments Adapted to the Use of American Surveyors.* 7th ed. Philadelphia, 1796.

Gunter, Edmund. *A canon of triangles: or a table of artificial sines, tangents, and secants, drawne from the logarithmes of the Lord of Merchistone.* London, 1620.

———. *Description and Use of the Sector, the Crosse-staffe, and Other Instruments.* London, 1624.

———. *The First Book of the Cross-Staffe.* London, 1620.

———. *New Projection of the Sphere.* London, 1623.

Hamilton, Alexander; Madison, James; and Jay, John. *The Federalist Papers.* Ed. Isaac Kramnick. New York: Penguin, 1985.

Hamy, E. *Joseph Dombey.* Paris, 1905.

Harris, Richard C. *The Seigneurial System in Early Canada.* Madison: University of Wisconsin Press, 1966.

Hart, John Fraser. *The Land That Feeds Us.* New York: Norton, 1991.

Hassler, Ferdinand R. *Coast Survey of the United States. [An answer to certain statements on that subject made in Congress.]* Philadelphia, 1842.

———. *Comparison of Weights and Measures of Length and Capacity.* Reported to the Senate of the United States, by the Treasury Department, Washington, D.C., 1832.

Hicks, Frederick C., ed. *Thomas Hutchins: A Topographical Description of Va, Pa, Md + No Carolina.* London, 1778; reprint, Cleveland: Burrows Bros., 1904.

Jacobs, Harvey M., ed. *Who Owns America? Social Conflict over Property Rights.* Madison: University of Wisconsin Press, 1998.

Jefferson, Thomas. *The Papers of Thomas Jefferson.* Ed. Julian P. Boyd. Princeton, N.J.: Princeton University Press, 1950–92.

Johnson, Hildegard Binder. *Order upon the Land: The US Rectangular Survey and the Upper Mississippi Country.* London: Oxford University Press, 1976.

———. *The orderly landscape: landscape tastes and the United States survey.* (Lecture) Minneapolis: University of Minnesota, 1977.

———. "The United States Land Survey as a Principle of Order." In *Pattern and Process: Research in Historical Geography.* Washington, D.C.: Howard University Press, 1975.

Keay, John. *The Great Arc.* New York: HarperCollins, 2000.

Klein, Arthur. *The World of Measurements.* London: Allen and Unwin, 1975.

Koch, Adrienne. *Jefferson and Madison: The Great Collaboration.* University Press of America, 1986.

Kula, Witold. *Measures and Men.* Trans. R. Szreter. Princeton, N.J.: Princeton University Press, 1986.

Leybourn, William. *The Compleat Surveyor, Containing the Whole Art of Surveying Land, etc.* London, 1653.

Livermore, Shaw. *Early American Land Companies and Their Influence, a Corporate Development.* Commonwealth Fund, 1939.

Love, John. *Geodaesia or the art of surveying and measuring of land made easie.* London, 1688.

Marryat, Frederick. *A Diary in America.* Ed. Jackman, Sydney. New York: Knopf, 1962.

Meinig, D. W. *The Shaping of America: A Geographical Perspective on 500 Years of History.* Vols. 1–3. New Haven, Conn.: Yale University Press, 1986, 1993, 1998.

Melish, John. *Travels in the United States 1806, 1807, and 1809.* Philadelphia, 1813.

Merrell, James. *Into the American woods: negotiators on the Pennsylvania frontier.* James H. Merrell. New York and London: Norton, 1999.

Nash, Roderick. *Wilderness and the American Mind.* New Haven, Conn.: Yale University Press, 1982.

National Institute of Standards and Technology. *Toward a Metric America.* Washington, D.C.: Department of Commerce, 1971.

Pasley, Jeffrey. "Private Access and Public Power." In *The House and the Senate in the 1790s: Petitioning, Lobbying, and Institutional Development.* Ohio University Press, 2002.

Pattison, W. D. 1957. *The Beginnings of the American Rectangular Land Survey System.* Ohio Historical Society, 1970.

———. *Beginnings of the American Rectangular Land Survey System, 1784–1800.* Research paper No. 50. University of Chicago Department of Geography. Chicago. 1957.

Price, Edward T. *Dividing the Land: Early American Beginnings of Our Private Property Mosaic.* Chicago: University of Chicago Press, 1995.

Putnam, Rufus. *Memoirs of Rufus Putnam.* Ed. Rowena Buell. Boston: Houghton Mifflin, 1904.

Pynchon, Thomas. *Mason and Dixon.* New York: Holt, 1997.

Raban, Jonathan. *Bad Land: An American Romance.* London: Picador, 1996.

Reps, John. *The Making of Urban America: A History of City Planning in the United States.* Princeton, N.J.: Princeton University Press, 1965.

Richeson, A. W. *English Land Measuring to 1880.* Cambridge, Mass.: MIT Press, 1966.

Sakolski, A. M. *The Great American Land Bubble: The Amazing Story of Land-Grabbing, Speculations, and Booms from Colonial Days to the Present Time.* New York and London: Harper and Brothers, 1932.

Schama, Simon. *Landscape and Memory.* New York: HarperCollins, 1995.

Selby, John. *The Conquest of the American West.* London: Allen and Unwin, 1975.

Smalley, E. V. "Isolation of Life on Prairie Farms." *Atlantic Monthly,* September 1893.

Smith, Thomas H. *The Mapping of Ohio.* Kent, Ohio: Kent State University Press, 1977.

Steele, A. R. *Flowers for the King. The Expedition of Ruiz and Pavon and the Flora of Peru.* Durham, N.C.: Duke Univeresity Press, 1964.

Stilgoe, John. *Borderland: Origins of the American Suburb, 1820–90.* New Haven, Conn.: Yale University Press, 1988.

———. *The Common Landscape of America, 1580–1845.* New Haven, Conn.: Yale University Press, 1982.

Taylor, E. G. R. *Mathematical Practitioners of Tudor and Stuart Period.* London, 1954.

Thompson, F. M. L. *Chartered Surveyors: The Growth of a Profession.* London: Routledge and Kegan Paul, 1968.

Trollope, Frances. *Domestic Manners of the Americans*. London: Imprint Society, 1969.

Turner, Frederick J. *The Frontier in American History*. New York and London: Dover, 1996.

U. S. Congress. Journals of the Continental Congress 1774–89. Ed. Worthington C. Ford et al. Washington, D.C.: 1904.

Wallace, Anthony F. C. *Jefferson and the Indians: The Tragic Fate of the First Americans*. Cambridge, Mass. and London: The Belknap Press of Harvard University Press, 1999.

White, C. Albert. *A History of the Rectangular Survey System*. Washington, D.C.: Government Printing Office, 1982.

Williams, Penry. *Life in Tudor England*. Batsford, 1964.

Wilson, Henry. *Geodesia catenea or surveying by the chain only*. London, 1786.

Winstanley, Gerrard. *A Declaration From The Poor oppressed People of England, Directed to all that call themselves, or are called Lords of Manors, through this Nation; That have begun to cut, or that through fear and covetousness, do intend to cut down the Woods and Trees that grow upon the Commons and Waste Land*. London, 1649.

———. *A Letter to The Lord Fairfax, And His Councell of War, With Divers Questions to the Lawyers, and Ministers: Proving it an undeniable Equity, That the common People ought to dig, plow, plant and dwell upon the Commons, without hiring them, or paying Rent to any*. London, 1649.

Wise, M. Norton, ed. *The Values of Precision*. Princeton, N.J.: Princeton University Press, 1995.

Worsop, Edward. *A discoverie of sundrie errours and faults daily committed by landemeters ignorant of arithrmeticke and geometrie*. London, 1582.

Wyman, Walker D. *Frontier Woman: the Life of a Woman Homesteader on the Dakota Frontier*. Madison: University of Wisconsin-River Falls Press, 1972.

Xenophon. *Oeconomicon*. Trans. Gentian Nervet. London, 1534.

Zengerle, Jason. "Waits and Measures." *Mother Jones*, January–February 1999.

Zupko, Ronald Edward. *Revolution in Measurement: Western European Weights and Measures Since the Age of Science*. Philadelphia: American Philosophical Society, 1990.

INDEX

boundaries, 33, 74
settlers, 36–37
Perch (a.k.a. rod, pole), 16, 26, 90, 142, 181, 251
Philadelphia, 34, 105, 134, 179, 181
Philadelpha Mint, 201, 202, 204
Philip II of Spain, 9
Physics, 55, 242–46
Picard, Jean, 68, 91, 125
Pickering, Timothy, 62, 148–49
Pied, 24, 90, 197, 198, 240
Pilgrims, 35
Plains Indians, 23
Plat(s), 9, 12, 18, 38, 73, 87, 156, 210, 235
depository for, 259
drawn to scale, 44
imaginary, 212
inaccurate, 40
Northwestern Territory surveys, 144
power of, 166, 168
registering, 191
unrealistic, 39
U.S., 191
Platinum, 131, 132, 239, 244
Plymouth Colony, 35, 44
Plymouth company, 31
Point of Beginning, The, 2f, 3, 6, 74
Pole, 10, 16, 142, 251
Polestar, 144, 192
Pollio, Vitruvius, 23
Pound, 68, 205, 253
imperial, 201
in metric system, 250
standardized, 237
Power, 244, 258
precision measurement as source of, 237
Powhatan Indians, 36, 44
Prairies, 225–29, 234, 235
weather, 225–26
Pratt, Orson, 182–83
Precision gauges, 237
Preemption Act of 1841, 166
Price(s), changing, 21–22
Property, 20, 90, 172, 174, 248, 251
colonists' idea of, 35
concept of, 170, 219, 221, 260–61

and democracy, 248
in France, 89–90
laws of, 92
legal problems of redefining, 260
Property rights, 44, 45, 48–49, 233
Proprietors, 34, 37, 38, 39, 47, 63
Proudhon, Pierre-Joseph, 90
Prussia, 241–42
Public land sales, 70, 71, 80, 81, 83, 149, 175
decimalized measures in, 113, 115–16
fraud and corruption in, 150–55
land size in, 164–65
minimum parcel, 166, 169
railroads, 166
weights and measures in, 141–42
Public lands, 233–34
distribution of, and society, 173–75
grazing rights, 235
Public lands survey, 3, 6, 110–11, 141–42, 144, 146–47, 155–56, 176, 179, 181, 192, 218, 220, 222–25, 259
administration of, 159
blueprint for, 76
California, 221–23
confusion of measures in, 197
and fraudulent claims in New Mexico and Arizona, 221
and landownership, 209
Louisiana Purchase, 158–59
measures used in, 115–16
pattern, 183
point of beginning, 3
reached Pacific, 209
speculation threat to, 148–50
and squatters, 163
suspended, 83, 88
Pulliam, John, 163–64
Pulliam, Robert, 164, 166–67, 168, 227, 229
Putnam, Rufus, 51–53, 52f, 65, 72, 76, 80, 82, 93, 156, 162, 163, 167, 193
acquiring land for veterans, 62–63
and ban on slavery, 172
death of, 148
and disposal of Western Territory, 73
and Jefferson, 89, 127
led settlers to Ohio Company land, 84–85

first in U.S., 236
industrial, 98
for kilogram, 239
for meter, 239, 242, 244–45
universal, 246–47
Standard measure, 108, 126
Standardization, 97–98, 219, 237
Star charts, 33
"Star" pound, 204, 205*f*
Star sightings, 33, 91
States
    boundaries, 225
    confusion of measures in, 197
    land sales, 149
    from Northwest Territory, 81–82
    paper currency, 80, 114, 149
    from Western Territory, 70
States, new, 174, 199, 237
    ban on slavery, 70, 82, 171
    shape of, 70–71
"Statute for Establishing Religious
    Freedom," 58
Steam engine, 98
Sterling system, 25
Stevin, Simon, 17
Stoute, Philemon, Jr., 227
Substance, 244
Suburbs, 186–87
Sugar Creek, 164, 167, 218, 226–27
Supreme Court, 154–55, 215
Surveying
    ancient method of, 8
    decimals in, 17
    difficulties of, in America, 32–33
    history of, 3
    Louisiana Purchase, 158–59
    method of, 12, 40–41
    metric system in, 259–60
    rectangles in, 220
    Southwestern Territory, 155–56
    Western Territory, 60, 70–71, 72–73
    see also Public lands survey
Surveying expedition, 77–79, 81, 83
    failure of, 84
Surveying manuals, 9, 13
    American, 21, 32–33

Surveyor-general, 34, 39, 155, 159, 160
    power of, 214
Surveyors, 7–8, 16, 141, 183
    advance men for speculators, 75, 79–80
    as agent of change, 12–13
    boundary commission, 74–75
    bribes to, 64, 70
    British America, 30, 31, 32, 38–39,
        45–46
    danger from Indians, 118
    deceit by, 212
    desire for accuracy, 99
    errors by, 162–63, 168
    fees, 38–39, 45, 79, 86, 155
    in France, 90
    grades of, 33
    grid, 166, 167–68
    Gunter's chain, 18
    international meter standard, 244
    and irregular shapes, 19
    maps by, 9, 10, 210
    Ohio, 145, 146
    professional standards, 159, 167–68
    survey parties, 86–87
    Western Territory, 73
Surveys, 12
    demand for, 39
    false, 261
    lands acquired from Indians, 211–13
    made land profitable, 38
    see also Public lands survey
Swanwick, John, 121, 139–40
Switzerland, 189, 190, 239, 241
Symmes, John Cleves, 79, 87, 145, 163
Système Internationale de Poids et
    Mesures, Le (SI), 246
Système usuelle, 240

Tallyrand-Périgord, Charles Maurice de,
    100–101, 102, 109, 121, 125–26
    and Louisiana Purchase, 156–57
Technology, 18, 253
Telescope, 33, 96, 191
Telegraphy, 237
Temperature, 244
Tennessee, 51, 63, 103–4, 154, 155, 214